Possibilist Knowledge

Possibilist Knowledge

A Rigorous Framework for an Indeterminist Science

Gustavo Castañon

BLOOMSBURY ACADEMIC
NEW YORK • LONDON • OXFORD • NEW DELHI • SYDNEY

BLOOMSBURY ACADEMIC
Bloomsbury Publishing Inc, 1359 Broadway, New York, NY 10018, USA
Bloomsbury Publishing Plc, 50 Bedford Square, London, WC1B 3DP, UK
Bloomsbury Publishing Ireland, 29 Earlsfort Terrace, Dublin 2, D02 AY28, Ireland

BLOOMSBURY, BLOOMSBURY ACADEMIC and the Diana logo are trademarks of Bloomsbury Publishing Plc

First published in the United States of America 2026

A catalog record for this book is available from the Library of Congress.

ISBN: HB: 978-1-6669-5396-1
ePDF: 979-8-216-25967-1
eBook: 978-1-9787-6547-4

Typeset by Newgen KnowledgeWorks Pvt. Ltd., Chennai, India

For product safety related questions contact productsafety@bloomsbury.com.

To find out more about our authors and books visit www.bloomsbury.com and sign up for our newsletters.

I dedicate this book to my daughter Helena and her future of infinite possibilities

"Everything that is possible demands to exist."

Leibniz

Contents

Preface

This book, which I now present for public evaluation, is ambitious yet, I hope, not naïve. The definitions and arguments within have undergone rigorous criticism over the past decade. Some may still be flawed, but it is up to new critics to show where. Even if not every ambition outlined in the Introduction were fully met, I believe that, by the end of this book, I will have provided new arguments for consideration and highlighted essential aspects of the problems of laws and explanations in science that have been neglected.

Parts of this book have already been published. Sections of Chapters 2 and 3 appeared in *Erkenntnis* under the title "Possibilist Explanation: Explaining How-Possibly through Laws" (Castañon, 2021). Additionally, material from my 2021 doctoral thesis (Castañon, 2021b) is spread throughout Chapters 3 and 4.

That's why I am grateful to CAPES (*Coordenação de Aperfeiçoamento de Pessoal de Nível Superior*), the Brazilian funding agency that supported the beginning of my research on this topic ten years ago. I must also thank my department at UFJF (Federal University of Juiz de Fora) for granting me that sabbatical year ten years ago and the present semester. My gratitude and appreciation to CHESS (Centre for Humanities Engaging Science and Society) at Durham University, where I started this research, and to the Philosophy Department of Harvard University, where I completed the investigation and the manuscript during my period as visiting scholar.

Special thanks to the staff at Bloomsbury and Rowman & Littlefield, who invited me to write this book and will publish and distribute it. I am particularly grateful to Deanna Biondi and Jana Hodges-Kluck for their patience throughout the revising and editing process.

I am especially grateful, truly, to the anonymous referees of *Erkenntnis* and Rowman & Littlefield, whose advice and critique really made this work much better. I also extend my gratitude to numerous individuals who have helped shape these ideas throughout the years, such as Nathalie Cadena, Andrew Pinsent, José Olavo Brando, Marco Rufino, David Papineau, Rune Nyrup, Fernando Rodrigues, Anna Alexandrova, Julian Reiss, Rodrigo Gouveia, Alberto Oliva, Mangabeira Unger, and Edward Hall. It is not rhetorical to say that this work would be much worse without their criticism. They were all contributory conditions to it.

Above all, I am particularly indebted to Pedro Merlussi and Nancy Cartwright for their sharp minds, invaluable insights, and extensive discussions on the matters outlined here. These two were necessary conditions for this work.

Introduction

We humans not only know some things that are physically necessary or probable; we also know some things that are possible or impossible to occur. I call this "possibilist knowledge." However, modern science consciously investigates and pursues only the first two kinds of knowledge. This book is about why this is the case, how to seek and utilize possibilist knowledge rationally, and what its impact is on our conception of science.

Possibilist knowledge is the knowledge of what is possible or impossible within a given context based on the satisfaction or absence of necessary conditions. Unlike knowledge about physical necessity, which tells us what must occur, and probabilistic knowledge, which tells us what is likely to occur, possibilist knowledge focuses on what could or could not happen under certain conditions. For example, knowing that combustion cannot happen without oxygen is a piece of possibilist knowledge, as it defines a boundary of possibility given a specific condition.

This book's primary contribution lies in the presentation, definition, formalization, and exemplification of possibilist laws and explanations in science, solving some old problems that have prevented them from being scientifically recognized or widely used. Moreover, it demonstrates their ubiquity, even though almost nobody is actively seeking them today.

However, there are also secondary implications. The concept of possibilist knowledge may also lead to a series of other significant consequences, some of them within the philosophy of science.

An enhanced possibilist knowledge will deepen the dissolution of the traditional association between scientific rigor and Laplacian determinism, showing how science can remain entirely exact without assuming a deterministic universe at all.

Across the sciences, possibilist tools refine our understanding of causation by distinguishing between conditions that merely permit an outcome and those that are authentically necessary for it, enabling investigators to target interventions more precisely. Additionally, by isolating a single veto condition to explain why an outcome could not occur, it cuts through multicausality, rehabilitates "negative causation," and recasts the deductive-nomological model (DNM) as a genuinely causal tool that illuminates mechanisms and their consequences.

In psychology, these epistemological and methodological improvements frame variables as causal pressures and vetoes rather than inexorable deterministic chains. It will strengthen the reliability of our predictions while still leaving room for agents to initiate new causal chains, thereby reconciling rigorous scientific practice with the everyday and legal presumption of free will. Such a change would vindicate a robust, empirically grounded notion of human freedom.

These improvements in our understanding of scientific knowledge, especially in psychology and neuroscience, could have a significant cultural impact, depending on their acceptance. From the Greek impasse between Democritean atoms and Stoic internals to today's neuro-deterministic headlines proclaiming that we are nothing but cerebral automata, Western culture has oscillated between the fear that freedom is an illusion and the hope that agency is real, which sustains our moral and legal concepts. This "ancient war" has resurfaced today with naïve allegations (supposedly based on functional magnetic resonance imaging [fMRI] studies) that premotor brain activations determine every human choice. Such allegations underpin claims that criminal responsibility, moral choice, and even creativity are vestiges of prescientific myths. An awareness that most scientific laws are possibilist could reshape neuroscientific claims, restoring a conception of human action that is empirically testable yet irreducible to deterministic schemes.

However, I will not address any political or social issues here. What I am doing, besides presenting possibilist knowledge (with rectification to the formalization presented in *Erkenntnis*), is offering new arguments about the nature of scientific explanation and scientific psychology. Other implications of possibilism addressed by this book are the rectification of the definition of "How-possibly explanations," a new definition of "mechanism," a restatement of the DNM of explanation in a causal form, a new approach to the problem of the relation between free will and laws of nature, and a new proposal for the limit between philosophical and scientific psychology, as well as a new incipient roadmap for this discipline unification. Well, it is undoubtedly very pretentious, I am aware. However, now it is up to the new critics to judge if I could offer some contribution to these themes.

A Brief Outline

Let me present some of the problems addressed by this book in the same order as the chapters. Chapter 1 introduces the idea of possibilism, distinguishing between different forms of possibility and necessity, and briefly justifies its name, also listing what is new in my formulation. It will also briefly defend the definition of science I will adopt in this book, among other features, as the activity that primarily predicts (the necessity, probability, or possibility of) events in the world.

Chapter 2 will define "possibilist law" as a type of scientific law that provides that, in the absence of a necessary condition N, event E is not possible. A possibilist law is not only an actual strict law (I will argue here that it is the only kind of actual strict law in the universe), but it is falsifiable by one and only one reliable observation of an event E in the absence of N. The needlessness for experiments to falsify it is a distinction that makes it especially suitable for the social sciences.

The possibilist explanation is the other central tool of possibilism, examined in Chapter 3. A possibilist explanation explains an event, showing how it was epistemically possible rather than why it would be epistemically necessary. Its explanandum is the epistemic possibility of an actual event previously considered epistemically impossible or improbable. A new concept is also introduced to define a possibilist explanation: permissive condition. A permissive condition P for an event E is something that entails a necessary condition N for it (but does not entail the event itself). Another problem addressed by the chapter in light of possibilism is the dispute between causal and nomological explanations. Possibilist explanations show William Dray's (1968) idea that how-possibly explanations do not need laws to explain, an illusion. It is shown that a how-possibly explanation is better executed by a DNM structure and reveals the implicit explanatory role played by laws in causal explanations. Following this, I argue that an accurate and complete scientific explanation should be causal and law-based.

In Chapter 4, I will also offer a new definition of "mechanism" as a function to better construct my argument. This definition will restrict the class of phenomena we could classify as mechanisms but enable it to explain other phenomena (capacities) precisely. It will restrict the scope but add knowability to the mechanistic explanation, which extensively uses possibilist knowledge. I will also argue for the compatibility of possibilism with mechanistic explanation and how it illuminates the concept of "negative causation." In doing this, the concept of "preventive condition" will be introduced.

A third goal of this book is to achieve something that was not in the original paper or doctoral thesis from which it originated: to provide a more comprehensive sample of possibilist laws and explanations that can be deduced from the various disciplines of modern science. It will begin with Chapter 5, where I list possibilist laws and explanations in physics, astronomy, chemistry, biology, history, geography, sociology, political science, and economics.

The sample of possible laws and explanations in psychology and neuroscience is in Chapter 6. Psychology is chosen as a detailed case study of the possible impact of possibilism in science because it is expected to have more significant implications, and also because I am more familiar with it. Today, this discipline remains in a pre-paradigmatic state, with neither the nature of its laws—if any—nor the boundary between its scientific and philosophical forms of inquiry has been established. In this book, I defend the idea that we can expect only two forms of laws from scientific psychology: possibilist and comparative ceteris paribus laws. By defending this, I argue that the limits of scientific psychology are laws that predict impossibility or causal pressures and that every prediction or explanation that sufficiently predicts or explains a psychological phenomenon is philosophical and unfalsifiable.

That leads us directly to a significant problem in the philosophy of psychology: the problem of the unity of discipline. An eventual widespread adoption of these theses about psychological laws and the limits of scientific psychology could confine the historical conflict and fragmentation in psychology to the field of speculation about the sufficient causes of a behavior. It would reintegrate theorists who accept the possibility of free will into the cognitive sciences. The unification of the discipline would occur in the field of evidence for necessary conditions and causal pressures for certain behaviors; however, it would never occur in sufficient conditions for any behavior.

Another implication, not only for psychology (including neuroscience) but also for all social sciences, is to the problem of multicausality. Given its intricate causal environment, possibilist knowledge is especially suited for these disciplines as it operates by establishing only one necessary condition for an event to be possible. Possibilism in psychology and sociology would offer explanations with a simple and precise format since it would be enough, at the limit, to determine a single variable to explain our explanandum: "As x was not present, given the law X, then behavior y was not possible."

I will finish this book with Chapter 7 by reflecting on some metaphysical consequences of possibilist knowledge. Suppose that, in psychology, scientific laws only predict causal pressures (probability) and vetoes (impossibility) to some behavior in some situation. In that case, they do not commit themselves to a Laplacian deterministic conception of human behavior. In other words, I will argue here that possibilist knowledge (as well as comparative ceteris paribus laws) is compatible with beliefs in incompatibilism, free will, and a universe regulated by strict laws altogether.

Final First Thoughts

From my earliest encounters with philosophy and cognitive sciences, the conflict between determinism and free will has become one of my primary concerns. It was one of my deepest motivations for continuing to reflect on the problems of this book since my undergraduate studies.

For now, I affirm that possibilist laws and explanations are legitimate and neglected parts of scientific knowledge and are especially suitable for the human sciences, the key to their longed-for scientificity. Moreover, possibilism compatibilizes modern science with the idea of free will without compromising it metaphysically. By the end of this book, I will have demonstrated these claims.

I am deeply convinced that we cannot abandon either of these civilizational treasures: the legal and political system built on the concept of individual freedom and the modern science that has propelled us to unprecedented heights of knowledge. Both are inherently valuable, the former morally right and the latter epistemically efficient. If you share these convictions, or at least these hopes, grant me the privilege of keeping your attention, your critical thought, and your mind open throughout this book.

1

Possibilism

Here is introduced the idea of possibilism. The first subitem presents it, illustrated by examples and intuitive definitions. After that, two clarifications: First, it is set that the kind of possibility possibilism deals with is primarily epistemic. Second, I justify my choice of the terms "possibilism" and "possibilist" to represent this modality of knowledge, briefly reporting its history in philosophy. Then, I discriminate what is new about this book's definitions and arguments on it. Finally, a working definition of modern science is settled, taking the opportunity to present an argument refuting the idea that Laplacian determinism is a necessary condition for science and defending that only a soft form of object stability is a necessary condition of it.

1.1 Possibilism, the Very Idea

Imagine you are at the movies. On the screen, the film begins with a scene of a man in his sixties seated on a bench in a park, flanked by a thirty-year-old man in a wheelchair. They are tense and sad. A dialogue begins. From it, you perceive that it is a father and his thirty-year-old son, called Steve. The father is saying, consternated, that it is past time for Steve to accept that he will never be able to walk again because his spinal cord is severed. The son cries and asks the father to go away. The crying father leaves. Steve watches, with an expression of pity, as his father moves away. A few seconds later, a woman on the side bench also leaves and accidentally drops her wallet. After a while, Steve calmly looks around, checks that nobody is watching, and stands up easily, walks to the wallet, takes it, and returns to his wheelchair.

What just happened? What kind of explanation are you and the rest of the audience expecting? You are not expecting someone to explain why he is a thief, nor that neuronal signals from his motor cortex have caused his legs to move up to the wallet. You do not want to know the reasons for his getting up, not even about the cause of his walking. What you want to know is *how it was possible* since, supposedly, he had a spinal cord severed.

Let us say it the other way. The explanation you demand in the face of such a situation is not about what *sufficiently caused the walk.* If I tell you that neuronal signals from his brain reached his legs, moving them, it will not satisfy you because you already know that it happened, somehow. What you want to know is *how* it happened, *how it was possible* for it to happen since, supposedly, it was epistemically impossible (i.e., it contradicted what we know).

Thinking about it, a lot of putative theories start to pop out. Maybe his father was deceived, and Steve never became a paraplegic. Or maybe this Steve is Steve Austin, a six-million-dollar human cyborg, and he had a new, groundbreaking technology installed in his spine, but he cannot tell his father because now he is a spy. There are a lot of "maybes," but all of them look to rebut the supposed impossibility of this specific situation.

That kind of explanation we are looking for is a possibilist explanation (PE). Possibilist explanations are demanded whenever an event E appears surprising or even impossible, under current scientific laws and known facts (or simply commonsense expectations). A possibilist explanation removes the surprise by identifying what I call a possibilist law (PL), which states a necessary condition N (connection between motor cortex and legs' neurons) for E (to walk), and a permissive condition that is a specific, contingent circumstance P (new technological device in spinal cord) sufficient to realize the necessary condition N (connection) for the occurrence of E (Steve walking). Once N and P are made explicit, the occurrence of the event is no longer puzzling: the necessary condition N shows what must hold for Steve walking to be possible (and apparently did not), while the permissive condition P shows that it did hold in the situation at hand.

Now, let us imagine another situation. A worried mother looks for a psychiatrist. She has a history of schizophrenia in her family. She wants the power of prediction and control that modern science offers over nature. However, she is not looking for medicine, a psychotropic that causes biological alterations in her newborn son. She is asking the doctor if there is a way to prevent an alteration in her son's brain in the future. If there is a known way to find out if her son has the biological marker of schizophrenia, and if it is the case, if medicine knows the way to make it go away, making it impossible for her son to develop this horrible illness.

Pay attention here. The woman wants a prediction. However, she does not want to predict how to sufficiently cause an alteration in the world, not even how to raise or diminish the probability of such an alteration. She wants to prevent a specific alteration, to make it impossible for something to happen.

She will get what she wants with a possibilist law. Instead of a prediction about what will necessarily or probably happen when some condition holds, she demands a prediction about what cannot happen when that condition is absent. This law, stated in the negative form, "if not x, then y cannot happen," is a possibilist law. It makes genuine prediction and falsification of possibilities feasible.

These two imaginary cases illustrate familiar cases of requests for knowledge (one explanatory, the other predictive) about possibility or impossibility rather than sufficiency or probability. That is precisely what I call "possibilist knowledge": knowledge about what is possible or impossible in nature.

Of course, I am not saying that science does not already address these questions. What I am saying is that it could do so much better—more extensively, more consciously, and more logically. Science has yet to tap the full potential of possibilist knowledge because it does not properly understand the logic underlying explanations of possibility and predictions of impossibility. Worse still, the prevailing metaphysics of science does not consider the possibility that the world, at its most fundamental level, operates in a possibilist way. This book begins to remedy that omission. In the chapters that follow, we will examine possibilist explanations and possibilist laws in greater detail, but first, we will take a brief detour to clarify the terms "possibility" and "possibilism."

1.2 Which Possibility?

Two parentheses here are appropriate. The first one is about what kind of possibility we are talking about. This is not a "scholastic" or marginal issue. When we use the terms "possibility" and "necessity," we may be meaning at least seven different things: logical, analytic, metaphysical, physical, epistemic, doxastic and deontic modalities.

To possibilist knowledge, epistemic and physical possibilities are what matter. When using a possibilist law, we want to represent a physical possibility (law of nature) but express a rule about epistemic possibility (scientific law). When giving a possibilist explanation, we explain an epistemic possibility to rebut a doxastic presumption (mere belief) of impossibility.

Distinguishing these kinds of possibilities from their neighbors—such as epistemic and doxastic possibility—helps avoid some confusion when we begin to consider possibilist laws and explanations. Here, I will distinguish logical, metaphysical, physical, epistemic, and doxastic possibilities, leaving analytic and deontic behind.[1]

Let us begin with *logical possibility*. To say that some proposition is logically possible is to say it does not entail a contradiction. Correspondingly, to say that some proposition is logically necessary is to say that it cannot be denied without contradiction. Using the possible worlds semantics, we can say that a logically necessary proposition is true in all possible worlds due to its logical form and the logical operators' meanings. In its turn, a logically possible proposition may be true in some worlds and false in others due to its logical form, not to the state of beings in the language or the world. It is logically contingent. Exemplifying, "If P, then no Q, we have no P, then Q is possible" ($\neg P \vee \neg Q; \neg P \rightarrow Q$) is a logically possible proposition. It does not entail contradiction, and its truth depends on the value of the variables and the structure of the proposition. It is not a logically necessary truth either because, in some possible world, we may have while "If W, then no Q," and "W is true." On the other hand, we have "If P, then P" as a logically necessary proposition. None can deny it without entailing a contradiction. In other words, there are no possible worlds where we can have P without P, disregarding what P is.

The second is the *metaphysical possibility*. A proposition is metaphysically possible regarding the nature of things that it expresses, not the sentence's structure or the words' definitions. It is a level below generality about logical possibility since everything that is a metaphysical possibility must be a logical possibility, but not vice versa. A proposition is metaphysically possible if it could be true in at least one possible world, considering the essences or natures of things, but not necessarily confined to existing physical laws. For example, it is metaphysically possible that the gravitational constant was the number *phi*, but not physically nor epistemically possible.[2] Conversely, a proposition is metaphysically necessary if it is true in all possible worlds because of the essence of the things it represents, not just its logical structure. For example, "Gold has atomic number 79." The difference between analytic and metaphysical is that analytic depends solely on meanings, whereas metaphysical necessity depends on the very nature of things beyond human knowledge or definitions.

Third, we have *physical possibility*. A proposition is physically possible if it does not contradict the real laws of nature (that we may or may not know). Exemplifying, we have "Today it is 40 degrees Celsius" as a physically possible proposition. Physically possible propositions are contingent and do not entail any contradiction, and their truth depends on the state in which things are in some specific situation. A proposition is physically possible if it is true in at least one possible world with the same laws of nature as ours. Conversely, a physical necessity is something that must be true given the real (not necessarily the known) laws of nature. In other words, every law of nature is a physical necessity. A proposition is physically necessary if true in all possible worlds with the same laws of nature as ours.

Fourth, we have *epistemic possibility*. A proposition is epistemically possible if it does not contradict our present stage of knowledge of the world. Let us differentiate it from physical possibility with one example. "Interstellar trips beyond light speed are impossible." This is an epistemic necessity given our current knowledge, but we are unsure if it is also a physical truth. A proposition is epistemically possible if it is true in at least one possible world where people have the same knowledge as ours. On the other hand, an epistemic necessity must be true given our knowledge of the world, as "water at the sea level and natural pressure conditions boils at 100 degrees Celsius." A proposition is epistemically necessary if true in all possible worlds where people have the same knowledge as ours. The difference between physical and epistemic possibility is tricky. One way to distinguish between them is always to consider that epistemic truths are about what we think we know nature is, and physical truths are about what it really is, disregarding our knowledge.

Fifth, and finally for our proposals, we have doxastic possibility. A proposition is doxastically possible if it aligns with our beliefs, meaning it does not contradict what we believe to be true. Correspondingly, a proposition is doxastically necessary if it must be true given our set of beliefs. It is entirely subjective. Doxastic possibility depends on an individual's beliefs, which may or may not be compatible with reality or even our current knowledge. An example is the proposition "There may be reptilians living among us." To those who believe in the existence of reptilian aliens, it is doxastically possible. It does not matter if there is no evidence to support such a claim. On the other hand, the proposition "The sun always rises in the east" is doxastically necessary in relation to our common sense. The fundamental distinction here is that doxastic possibilities and necessities depend on belief systems rather than empirical knowledge or physical reality. They reflect what we think is possible or necessary, not what is verifiable or real.

Possibilist knowledge as defined here is not about logic, metaphysics, or language. It is about our knowledge of the world. Every time we enunciate a corroborated possibilist law, we are expressing an epistemic necessity (scientific law) that aims to capture an underlying physical necessity (law of nature). Every time we give a possibilist explanation, we are trying to show that an event misjudged epistemically impossible was, in fact, epistemically possible. Failure to supply such an explanation may indicate that a putative scientific law, though supposedly epistemically necessary, is physically false.

There is often confusion in philosophy regarding the various uses of "possibility," particularly between the adjacent in my exposition sequence. It is not uncommon for professional philosophers to conflate logical and analytic, physical and metaphysical, or physical and epistemic possibilities. However, we should strive to avoid this confusion and let us not blur the lines between epistemic and physical possibility with the logical, analytic, or metaphysical use of the terms. In this book, these terms are always used in their epistemic sense concretely, aiming at their physical sense normatively. Hence, I use "scientific law" to denote an epistemic necessity and "law of nature" to signify a physical necessity. In other words, every scientific law strives to represent a real law of nature.

1.3 What about the Term "Possibilism"?

The second digression concerns the term itself. I should justify why I am choosing "possibilism" to express the kind of knowledge I am looking at. First, naming these kinds of explanations and laws helps draw attention to this branch of knowledge, which is currently neglected. Second, "possibilism" and "possibilist" seemed to me the best option to name it. The term "how-possibly explanation" strikes me as odd, inapplicable to laws and predictions, and currently tied to quite different notions such as merely possible explanations or speculative models.

The root "possible" comes from the Latin *possibilis*. The English suffix "-ism" is used to form nouns that describe a specific system or philosophy, usually, an ideology (such as "liberalism" or "communism"). Of course, that ideological sense is not intended here. I use the term to denote a modality of knowledge, an epistemic approach to possibilities.

The variation with the suffix "-ist," in "possibilist," generally indicates someone or something related to or specialized in something. In our case, "possibilist" is everything related to or realizing the modality of possibility. When Kant outlined twelve categories in his *Critique of Pure Reason*, he divided them into four groups of three, corresponding to the logical functions of judgment: Quantity, Quality, Relation, and Modality. Possibility (with Actuality and Necessity) was one of the three categories of modality. However, when naming the logical kind of judgment related to the category of possibility, Kant gave it the problematic name "problematic." Problematic judgments to Kant affirm the possibility of something without asserting its actuality—for example, "The leaf may be green." This kind of judgment does not assert truth or falsity but only the possibility of a proposition (which can be any kind of possibility). Possibilist knowledge is knowledge about these kinds of propositions and uses them to express epistemic possibilities (aiming at real physical possibilities).

I think the word "possibilist" is not only better to name this kind of judgment, as it is the best name in our occidental languages for the kinds of laws, predictions, and explanations I am going to extensively explain, formalize, and exemplify here, as I said, to represent an entire modality of knowledge.

As far as I have been able to learn, the term "possibilism" has three known uses in the history of philosophy. The most famous is Alexius Meinong's metaphysical theory of objects, which includes both existent and nonexistent entities. He presented his metaphysical thesis in the paper *Über Gegenstandstheorie*, published in 1904 in the volume *Untersuchungen zur Gegenstandstheorie und Psychologie*. It distinguishes between objects that exist in reality and objects that exist only as possibilities that can be thought about. To Meinong, objects can possess properties independent of their existence in reality. For example, a golden mountain or a round square may not exist, but they can still be thought about and described as having determinate characteristics. This theory rests on the distinction between *Sosein* (being such-and-such) and *Sein* (actual existence). The fact that an object does not exist in the actual world does not make it not have a nature, nor prevent it from being the subject of true propositions.

In other words, all conceivable objects have independent properties in relation to their existence in reality. Thus, an object does not need to be real to possess characteristics; it can merely be a possibility to have attributes. There is the manifestation or non-manifestation of these attributes in reality that makes or does not make this object real. Therefore, a nonreal but possible dragon can be discussed in terms of its properties (having wings, breathing fire) without those properties implying that it is real.

This view has sometimes been interpreted as an early form of modal realism insofar as it treats merely possible or impossible objects as legitimate constituents of ontology. Meinong did not work with the concept of possible worlds, which was developed much later by Kripke and Lewis. However, his theory was the first to provide a way for thinking about the ontological status of non-actual entities. And it did it long before the dawn of the semantics of modal logic. However, we should be aware of the difference between Meinong's view and contemporary modal realism, particularly in its treatment of impossible or contradictory objects. We may say that in Meinong, possible worlds or scenarios are considered just as existent as the actual world, but not as real as it is. By contrast, Lewis's "possibilism" (Lewis, 1986), which he contrasts with "actualism," holds that merely possible objects are just as real as actual ones; they merely reside in other possible worlds. I should stress, however, that my use of the term "possibilism" in this book is strictly epistemic and thus distinct from Meinong's metaphysical thesis.

The oldest use of the term "possibilism" was made by the French geographer Paul Vidal de La Blache, also at the end of the nineteenth century. He is regarded as one of the founders of modern geography, mainly known for the concept of *genre de vie* (mode of life) and his geographic possibilism.

Opposing the deterministic interpretation of human culture, which posits that the natural environment (particularly the climate) sufficiently determines all its aspects, La Blache (1967) argued that the natural environment only imposes conditions on how culture can develop; it does not determine its final form. In other words, it restricted their possibilities for development but did not sufficiently cause its form.

A famous saying that would characterize the approach, according to Johnston et al. (2009), would be that of Lucien Febvre (1932, p. 27), according to which "There are no needs, but possibilities everywhere, and the human being, as master of possibilities, is the judge of their use." Arguing in favor of human agency and autonomy, La Blache emphasized the role of human free will in shaping cultural outcomes. For him, while the environment sets certain constraints, it does not dictate human activity. According to his possibilism, the environment restricts the possibilities to a specific range for the people living there. However, human agency will use these possibilities through different choices and even create new possibilities through the creative development of new technologies.

According to Johnston et al. (2009), possibilism should be interpreted as an ideal type, not as an operational research perspective. Despite being poorly developed from a logical and philosophical perspective, this geographical perspective is similar to some of the theses I will defend, and some of its findings will be presented as examples of possibilist knowledge here. Geographic possibilism should be understood as a methodological orientation rather than a metaphysical claim: It highlights the plurality of viable paths open to human societies within given physical constraints, rather than proposing a deterministic or predictive theory of social development. In turn, what I call here "possibilism" describes forms of laws and explanations that exist, whether we are aware of them or not.

The third contemporary use of "possibilism" concerns an ethical thesis by Patricia Greenspan (1978), which is distinct from the geographical concept of possibilism discussed earlier. According to this thesis, an agent is obliged to perform an act only if it is part of the ideal set of best acts they can perform throughout their life. Timmerman and Cohen (2020) define ethical possibilism as the thesis that "at t an agent S is obliged to ϕ at t′ if and only if S can ϕ at t′ and ϕ at t′ is part of the best set of acts that S can perform from t′ until the last moment in which she can perform an act."

Greenspan's possibilism opposes another ethical position known as actualism. Actualism focuses on what agents will do in the future based on their intentions or likely actions. It argues that the moral permissibility of an action should depend primarily on the specific future that the agent is likely to bring about.

In its turn, ethical possibilism addresses how individuals can be morally responsible for decisions when the outcomes of those decisions are uncertain. That contrasts with actualism, which emphasizes certainty or predictability in moral judgments. Greenspan argues that moral responsibility does not require certainty about what will come from our actions, but only reasonable judgments of it, based on available information.

Her "possibilism" is the thesis that, when evaluating the moral permissibility of an action, we should consider all the possible future scenarios that might follow from the agent's choices, not only the ones that are likely to occur given the agent's intentions.

This theory emphasizes the range of possible morally significant outcomes, recognizing that different decisions can be equally moral in uncertain consequences.

There is no direct relationship between ethical possibilism and possibilism as I define it here. Nevertheless, one of the reviewers of this book noted that, as ethical possibilism relies on the possible outcomes of specific actions and the agent's understanding of potential causal chains, possibilism and ethical possibilism may be closely linked. This link, however, will not be developed or studied here.

1.4 What Is New about Possibilism

Explanations about possibilities and predictions using necessary conditions have existed since humanity developed language. So, what is being proposed as "possibilism" for the first time?

First, its acknowledgment as another form of knowledge, distinct from sufficient and probabilistic. It is done by showing:

1) the logical structure of a possibilist explanation, or an explanation about the possibility of something given its apparently epistemic impossibility (as summarized in Section 1.1 and formalized in Section 3.5),
2) that this kind of explanation depends on an implicitly assumed possibilist law, which predicts the impossibility of an event in the absence of its necessary condition. That refutes the thesis of William Dray (1968) that "How-possibly explanations" (the predecessor of possibilist explanation) do not rely on laws. On showing these two things, I will defend
3) that possibilist explanations and laws can be scientific, based on empirical and falsifiable claims, and therefore, a new modality of science, alongside the modalities of necessity and probability.

A concrete illustration may help to preview it. Consider Ohm's law[3] in its possibilist reading: If a circuit is open (¬N), electrical current (E) is impossible (¬◊E). One reliable observation of current flowing across an open circuit would therefore falsify the law, whereas showing how the circuit was actually closed (or how a bypass restored continuity) would constitute a possibilist explanation of the apparently impossible current.

The process of achieving these three aims will go through other intermediate (and new) aims, offering:

- Refutation of Reiner's (1993) argument against scientific explanations of possibilities
- Solution to Dray's (1957, 1968) puzzles about "how-possibly explanations"
- Reductive analysis of possibilist explanation
- Definition and formalization of possibilist laws
- Definition and formalization of impossibilist laws

- Logical structure of the falsification of a possibilist prediction[4]
- Definition of permissive condition
- Definition of preventive condition

Let us begin working on this chapter by examining the problem of the scientificity of explanations, laws, and predictions of possibilities. Generally, the possibility of a scientific possibilist law is denied because possibilities are unfalsifiable, and the possibility of an explanation where the explanandum is the possibility of the occurrence of a phenomenon is equally denied.

To explain how they can be scientific, we should examine the implicit structure of a possibilist explanation more closely. Let us provide a clearer explanation of what we demand regarding Steve. Back to the movies, you would be satisfied if later in the film there were an explanation like this:

1) Having a neural connection between the motor cortex and the muscular nerves of the legs is a necessary condition for a voluntary walk.
2) There is a new technological device that realizes the neuronal connection between the motor cortex and legs.
3) When Steve Austin was recruited to become a spy, he received (T1) this new technological device in his spinal column; therefore:
4) His neuronal connection between his cortex and legs was restored; therefore:
5) His walking behavior was epistemically possible (in T2).

Note that the behavior was not epistemically necessary. It would be far more complicated (or impossible, if it involved a free choice) to explain scientifically why Steve walked and took the wallet than to explain how it was possible. Nevertheless, in this case, few people would even care about a deductive nomological or a causal explanation. Most people would accept good (and unscientific) psychological explanations for his motives in taking the wallet. Nobody is looking for a utopic deductive-nomological explanation (DNE) that, if feasible, would have to give the entire set of laws and conditions that regulate what has sufficiently caused the behavior.

Likewise, it is essential to point out the shared knowledge beyond the explanation. The audience (most of them) shares the knowledge that present technology cannot regenerate neurons, that a neuronal connection between the brain and legs is a condition for walking, and that Steve had his spinal column severed. That makes the fact that Steve walked to take the wallet require an explanation. And an explanation for the event might well have been that a new treatment had previously regenerated his neurons without his father's knowledge or that the woman was also a spy, and he and his supposed father had performed a pantomime because they were being monitored.

Anyway, what is to be rebutted in this kind of explanation is the presumption that a known natural law was violated (neuronal connection is necessary to walk), by showing that a necessary condition (neuronal connection) was restored (regenerated neurons) or had never gone (his spine was never severed).

Succeeding, such an explanation is possibilist—an explanation about the possibility or impossibility of an event. Moreover, the law to which we must recur to explain the event adequately is a possibilist law (1—Having a neural connection between the motor cortex and the muscular nerves of the legs is a necessary condition for a voluntary walk). This law predicts that a kind of event is impossible without a necessary condition.

A caveat is appropriate. Science certainly deals with the general properties of kinds of phenomena, but its social and practical power comes from using those generalities to predict and explain concrete, particular spatiotemporal events—tokens of those kinds. That is why science must draw on possibilist laws and explanations: laws that predict the impossibility of an event when its necessary condition is absent, and explanations that show how an apparently impossible event became possible. This is what I will demonstrate in this book.

A probabilistic formulation can, of course, express the same relations. For example, by assigning probability zero to an E-type phenomenon when its N-type condition is absent (the possibilist case) or by declaring that the conditional probability of B given A is one (the sufficiency case). Yet such uses do not change the fact that relations of sufficiency or possibility belong to a distinct—or, if one prefers, limit—modality that are more economically expressed, calculated, and understood in a sufficient or possibilist framework. A possibilist statement needs only the binary fact that an event is ruled out (or allowed) by a single necessary condition. It dispenses with the full probability distribution.

All possibilist explanations use possibilist laws, implicitly or explicitly. However, possibilist laws do not exist primarily to explain but to predict. Let us go back to the previous mother. What she is demanding from the doctor is some knowledge about a necessary condition for her son to develop schizophrenia. The objective here is to interfere with nature and remove such a necessary condition from her son.

For example, having the biological marker of schizophrenia is a necessary condition for developing it in the future. However, as we now know, it is not a sufficient condition for it. So, some other trigger—environmental, biological, chemical, or psychological—must appear at the right time (a developmental window) to activate the biological marker (a gene). Knowing which trigger can do this and when would make it possible for her to block it and prevent the development of unwanted symptoms in her child.

A possibilist explanation is a more potent form of "how-possibly explanation" (HPE), formalized into a covering law argument. HPE was proposed by William Dray (1954) as an alternative to the Deductive-nomological model (DNM) in the discipline of history. It stems from the fact that, sometimes, in ordinary life, explaining is just to set necessary causes, not sufficient ones: to show how something that seemed impossible or unexpected was physically possible, rather than showing why it was physically necessary (as we saw with Steve walking).

Dray believed this is a kind of explanation often used by historians, which makes no use of laws or generalizations. The fundamental difference between HPE and DNM lies in the type of question they answer. While the DNM works according to the pattern "Why-necessarily? Because...," the HPE works according to the pattern

"How-possibly? It could be like…."[5] It is usually used in history and evolutionary biology (Forber, 2010) to explain how it was possible that something predicted by a supposedly strict law or a ceteris paribus law did not happen.

1.5 Against Possibilist Science

This section outlines the standard objections to possibilist knowledge; their refutation is reserved for the following two chapters. HPE was criticized in some ways, but one difficulty is especially relevant here. Reiner (1993) demonstrated that pointing out particular conditions (such as the device reconnecting the spinal cord) as necessary is a mistake because, at best, they are considered separately as one of several conditions that can make the event possible (regenerated neurons, alien technology). I will refer to these kinds of conditions as "permissive." For him, a necessary condition N for the event E (if E then N) should be a set disjunctively formed listing all (permissive) conditions that could enable the event. However, it is impossible to establish definitively. Here is a summary of Reiner's argument against *the possibility of an explanation of a possibility*:

P1—To explain why an event was epistemically possible, it is necessary to establish the satisfaction of some necessary condition.
P2—Establishing the satisfaction of a necessary condition of a concrete event requires the exhaustive enumeration of all the particular conditions that disjunctively define[6] it.
P3—The particular conditions that disjunctively form a necessary condition are infinite.

Therefore,
C1—Scientifically explaining the possibility of an event is impossible.

In Chapter 3, I will refute Reiner's argument by proving that P2 is false and P3 is meaningless because of his error about the nature of the relevant conditions to this type of explanation.

However, this is not the only challenge to the possibility of a possibilist science. An apparently definitive veto comes from the fact that a prediction of a possibility is unfalsifiable. If we predict that if we have sun and rain at the same time, a rainbow is possible, whether a rainbow appears or not, the prediction "a rainbow is possible" is never contradicted. This is the structure of this argument:

P4—Every scientific prediction is, in principle, falsifiable.
P5—A prediction that an event E is epistemically possible since a necessary condition N is the case is not falsifiable.

Therefore,
C2—A possibilist law cannot be scientific.

I will also refute this argument, now in Chapter 2, by demonstrating that P5, the prediction of some impossibilities, is, in fact, falsifiable. Recalling from the last item that probability zero and impossibility belong to different modalities, conflating them is what makes the standard falsifiability worry look persuasive. With those two refutations, plus the specification and exemplification of possibilist laws and explanations, I will demonstrate that possibilist knowledge can be scientific. However, to do so, we should now make another detour and determine what we assume here by "scientific."

1.6 Which Is Science's Business?

An attempt to define the true nature of modern science is not among this book's pretentious objectives. Therefore, here I will only present the definition I assume of it to analyze the compatibility of possibilist knowledge with scientific knowledge. I do not feel obligated to engage in more contemporary discussions on the nature of scientific inquiry, including critiques and advancements in methodologies like Bayesian inference and simulation modeling. Moreover, this book is not a postmodern narrative with a new and obscure definition of science.

1.6.1 Definition of Modern Science

The activity that matters here is the one that emerged from the end of the Renaissance with the advent of the scientific revolution. From that moment on, "science" would no longer be the result of the privileged intuition of a wizard or the commentary of a philosopher of indisputable authority. Science begins to qualify as an autonomous body of knowledge because it seeks to test its hypotheses through experiments, which are publicly controllable and continuously improved by new and more precise measuring instruments that mathematize empirical experience.

The definition of science and the criteria of scientificity are still subjects of considerable dispute. I myself have spent much of my academic life discussing this. But not here. The present book has a different target: the existence and nature of possibilist laws and explanations. To avoid diving into a second theoretical quarrel, I shall begin with a simplified and idealized baseline: Ernest Nagel's definition in *The Structure of Science* (1961). In doing that, I recognize falsifiability (Popper, 2002/1952), truth (Niiniluoto, 1999), and explanatory unification (Kitcher, 1993) as normative ideals of science. It is also a deliberate methodological strategy: If possibilism can satisfy Nagel's standards, adapting it later to more pluralistic accounts will be straightforward.

I do not find substantive novelty in the interpretations of science proposed by Helen Longino (1990) and Ronald Giere (2006). Longino recycles social constructivism, interpreting objectivity as the product of critical interaction within diverse research communities, a view that foregrounds the social conditions of inquiry. My project, however, is concerned with truth-directed norms of empirical testing rather than with community-based standards of acceptability or "local epistemologies." Giere's "perspectival realism" likewise portrays science as the construction and coordination of partial, purpose-shaped models that track selected features of a mind-independent world. Nevertheless, its emphasis on perspectival reliability does not yield a single, comprehensive description of nature. Because the account developed here seeks broadly invariant, law-like, truth-oriented principles, these social-constructivist and perspectival frameworks are not considered.

By contrast, Nancy Cartwright's program (1983, 1999, 2021) offers a genuinely different conception to Nagel's: Science is the art of designing and deploying local nomological machines to predict, explain, and manipulate specific phenomena. To her, universal, exception-free laws are, at best, ideal templates; what is real are the capacities of things. The models illustrate how these capacities interact with specific circumstances. While I do not generally disagree with the metaphysical framework Cartwright (1999) offers, I adhere to the predictive and explanatory role of laws regarding fundamental entities, and I will argue for this position in the following three chapters, analyzing aspects of Cartwright's thinking.

For now, then, I stay with Nagel's classical six-point reconstruction as a methodological idealization: a sharply defined, law-oriented model that foregrounds the features of systematic lawful structures, formal clarity, and revisability. Treating it as a provisional baseline, rather than a final verdict on science, focuses the book on the modal questions of possibilism at hand and makes later departures theoretically salient. Pragmatically, a rigid starting point is useful: If possibilism holds under Nagel's stricter definition, its migration to looser, more pluralist settings should be easier.

In short, Nagel defines science as an activity with six essential characteristics:

1) A theoretical building with a set of consistent laws systemically organized.
2) Clarity of scientific laws and theories, established in formally impeccable and semantically univocal language.
3) Empirical research methods, which also establish the object of study and the relevant facts to study it.
4) Methods are controllable, reproducible, and intersubjectively observable.
5) The search to reduce phenomena to their deepest level of foundation.
6) Its knowledge is always open to revision and falsification.

Synthesizing these six characteristics into six words, we have that, to Nagel, modern science is systematic, rigorous, empirical, objective, reductive, and falsifiable. Science would be the rational investigation of the physical world through measured and controlled empirical testing of hypotheses. This approach allows for the approximation of universally valid knowledge in the form of general theories and universal laws, which predict, in an increasingly accurate and mathematical manner, empirical phenomena.

The business of science is predicting nature, our behavior, and the behavior of our devices. That is how it gains its enormous prestige and power in modern society. Science also offers explanations of the world and rigorous epistemic criteria to justify true beliefs, but, in practice, these are additional benefits of its enterprise.

1.6.2 Science's Presuppositions

Applying the transcendental method, we should also ask ourselves what needs to exist so that something like modern science can exist. The possibility that there is such a thing as a science that obtains secure knowledge in the form of laws of nature is not self-evident. The transcendental analysis yields the sine qua non conditions for the reality of such a science; it does not by itself demonstrate that they are in fact realized. It merely establishes that if science does exist in the sense just described, then those conditions must obtain. In other words, the conclusion is strictly conditional.

So, if we define "modern science" as above, it proves neither that it is correct from the point of view of the linguistic use of the term (descriptively) nor that it is the best description possible (prescriptively). It only implies we can deduce that, for it to be true, we must also assume the ontological assumptions:

1) There is a real world (realism).
2) We can obtain knowledge about it (epistemological optimism).
3) It submits its functioning and results to the laws of logic.
4) The universe exhibits at least some stable regularities.
5) The universe is at least in part mathematically describable.
6) We can represent reality through language (representationalism).

I am not going to work on all those ontological assumptions in our definition of science, only on the one that is essential to us here: the fourth transcendental condition, which states that the universe is at least in part regular, and therefore predictable.

The existence of an object in reality is not enough for it to be studied scientifically. Once we admit formulations of causal hypotheses as scientific knowledge, we must necessarily assume that the object being contemplated with these hypotheses, in at least some of its aspects, is regular (subjected to laws). If it had no regularities, science could not describe any stable characteristics of it; therefore, no knowledge would be possible about it.

Let us put it differently. One of the necessary attributes of scientific knowledge, as defined here, is that it consists of a set of propositions that allow us to predict or describe the physical world. For these propositions to be possible *a priori*, the universe must have invariable aspects; otherwise, not even a description of any part of it would be possible, let alone a prediction. I refer to this epistemological principle as the "principle of object regularity." We can only know about objects' regular aspects.

This problem is directly linked to Hume's concerns about the rationality of induction. He examined the assumption of nature's uniformity, arguing that our expectations that the future will resemble the past have no rational or empirical basis, even though we rely on them pragmatically. Proceeding on that trail, Popper defended that while the uniformity of nature is a necessary working assumption, it is always subject to testing and potential refutation. Both, therefore, assume that the regularity of nature is a necessary assumption for the quest for natural knowledge.

Regarding science specifically, the existence of empirically testable consequences of a theory does not require that sufficiency laws necessarily govern the phenomenon it intends to describe in all its aspects, but that it presents at least some regular aspect that can lead to predictions or descriptions. Even if this aspect is nothing more than the pattern by which this described object constantly changes (as Plato taught us). In other words, the existence of some physical impossibility in the universe is enough to establish some scientific knowledge about it: the knowledge about this impossibility.

1.6.3 Science and Determinism

So, scientific activity is characterized, in short, by the rational search for the discovery of the laws that govern phenomena. The belief in the regularity of the object is linked, in turn, to determinism and naturalism, which have formed the basis of modern science since Galileo Galilei. Naturalism is the belief in a causally closed universe governed by eternal laws, not subjected to magic, gods, chance, or chaos. See how Galileo (1623/1957) describes this belief in this classic passage:

> Philosophy [i.e., the study of the universe] is written in this grand book—I mean the universe—which stands continually open to our gaze, but it cannot be understood unless one first learns to comprehend the language and interpret the characters in which it is written. It is written in the language of mathematics. (p. 237)

In other words, Galileo has a deep belief that mathematical forms govern the world, that nature is stable and governed by mathematical laws, and therefore subject to discovery. For him, nature speaks the language of mathematics. It can therefore only be known through this language, that is, through questions that are correctly asked using a new method: experimentation, which applies the laws of measurement and mathematical interpretation to experience.

1.6.3.1 Definition and Differentiation of "Determinism"

Nevertheless, this mathematical conception of the universe should not be confused with the concept of "determinism." To understand it correctly, the first task is to define "determinism." Elsewhere (Castañon, 2021b), I defined it as "the thesis that antecedent conditions and the natural laws that govern them sufficiently cause every event." That means that given the universe's current state of being plus the entire set of actual laws of nature, things at any time in the future (and in the past) are already set and cannot be any other way. We often refer to this position as "nomological determinism" because the absence of alternative possibilities in the future is attributed to things governed by laws.

Another form of defining it was given by Karen Vihvelin (2013, p. 3), determinism would be defined by the conjunction of two propositions: (1) there are no exceptions to the laws that govern everything in the universe and (2) the laws of nature state sufficient, as opposed to merely necessary or probabilistic conditions. As we will see in

Chapter 2, Merlussi and Cartwright (2018) refer to these two domains as "extent" and "permissiveness."

We must carefully distinguish this nomological determinism from the other two less-known term uses. The first one is logical. What we call "logical determinism" (Jordan, 1963) is the idea that, logically, there is already a true proposition about whether I will write tonight; therefore, it seems that it is not up to me whether I will write or not. The second one is theological. According to "theological determinism" (Fischer & Tognazzini, 2013), since God would have omniscience and, therefore, knowledge of all true propositions about the future, it would not be up to us to make any choice about them. Both these other definitions of determinism are of no interest to us here and will not be referred to in this book anymore.

Therefore, from now on, I should only refer to nomological determinism as "determinism" or "Laplacian determinism." This position is commonly called this way because of its first formulator, the French scientist and philosopher Pierre-Simon Laplace. In his work from 1814, *Philosophical Essay on Probabilities* (1996/1814), he speculated that if an intellect (Laplace's Demon) knew the precise position and velocity of every particle in the universe at any given moment, plus the entire set of actual laws of nature, it would be able to predict both the past and the future of every particle perfectly. This is the idea that each state of the universe is entirely determined by its previous state.

Here, I question whether Laplacian determinism is required for science, not whether it is true. Its global claim differs sharply from the modest (and uncontroversial) observation that many local mechanistic systems (such as frictionless pendulums, sealed ideal gas chambers, digital logic gates) behave deterministically under tightly controlled conditions. I believe in such domains of local determinism within a natural landscape that also includes possibilist and probabilistic laws (of nature). However, the broader metaphysical truth of universal determinism lies beyond the present discussion.

1.6.3.2 Why Determinism Is Not Necessary to Science

Of course, Laplacian determinism is compatible with science, but it is not a necessary condition for it. For science to predict, describe, and obtain logical and falsifiable laws and descriptions, it only requires that at least some aspects of the object of study be stable, not all.

Let us refer to my preferred object of exemplification in the classroom: a chair. For a chair to be describable, it needs to have only one stable characteristic, like form. Suppose all the other characteristics, such as weight, height, color, and material, were to vary constantly in a semi-chaotic world. In that case, one single feature (say, form) remaining stable is sufficient to describe an object in terms of it.

Nagel (1961) defined modern science, among other things, as a theoretical building with a set of consistent laws systemically organized, and also as producing falsifiable knowledge. If these characteristics are true, for science to exist, it is necessary that the scientist believes in the laws of nature and searches for them. More than this, they must be stated in the form of laws or generalizations that precisely predict some state of facts in the world that we can test.

While the above points hold, I challenge the imposition of determinism as a transcendental condition for science. This metaphysical belief envisions a universal realm of necessity, where all relations in the universe are governed by sufficiency laws (or "sufficiency mechanisms"), whether known or unknown. A "sufficiency law" is our traditional *strict law*, which states sufficient conditions for an event to happen.

What is necessary for the emergence of modern science is that the universe is ruled by laws, not by sufficiency laws. The demonstration of the existence and fallibility of scientific possibilist laws would not prove that there are possibilist laws of nature. It would only prove that precise and falsifiable knowledge of natural laws could be derived from scientific sufficiency laws. The examples of contemporary quantum physics should have been sufficient to make this point, but now it is epistemically argued. Moreover, possibilist and probabilistic laws also demonstrate that, even if, as Nancy Cartwright (1983) defends, there are no such things as strict sufficiency laws in the universe, we can still conduct modern science.

The argument developed here demonstrates that rigorous, predictive science does not require determinism; hence, determinism is not a necessary condition for science. In Chapter 7, I revisit the problem of free will and determinism, argue against compatibilism, and maintain that any genuine reconciliation of science with free will must rest on a precise, falsifiable, and indeterministic science.

1.7 Final Remarks

This chapter introduced the concept of possibilism, laying the groundwork for a distinct modal form of scientific knowledge. By presenting intuitive examples and clarifying different kinds of possibilities—logical, metaphysical, physical, epistemic, and doxastic—it establishes a clear framework for possibilism within epistemic and physical possibilities.

The justification for using the terms "possibilism" and "possibilist" has been given, emphasizing the need for a distinct vocabulary to describe this neglected branch of knowledge. This chapter provided some historical context showing the precursors of the idea of possibilism, differentiating them from what I present here: the formalization and systematic exploration of possibilist laws and explanations, which seek to demonstrate that possibilist knowledge can be scientific. Additionally, it highlights some traditional criticisms regarding the possibility of scientific knowledge of possibilities, anticipating their refutation in the future.

Finally, I assume Nagel's traditional definition of science—systematic, rigorous, empirical, objective, reductive, and falsifiable—to demonstrate how possibilist knowledge aligns with these characteristics, thereby legitimating possibilist explanations and laws as legitimate scientific knowledge. From this definition, I have shown that only the principle of object's regularity is a necessary condition for science, therefore refuting the thesis that determinism is a necessary condition of modern science.

Now that we understand what I am calling "possibilism" and "science," let us refute the allegations against possibilist knowledge and analyze the models of laws and explanations presented in detail, thereby solidly establishing scientific possibilist knowledge.

2

Laws and Possibilism

Here, I will talk about possibilist laws (PLs) and the laws of nature. First, I will consider what laws are, explaining three central theses about it in contemporary metaphysics: Lewis's version of Hume's position, the "best system account," the thesis of laws as nomic necessity (necessary relations between universals), and the thesis of essentialist dispositionalism. Next, we will discuss the properties that a law of nature must possess. In the sequence, I will set the best representation available of laws and their basic logical form. We will examine the three principal forms of it, focusing on the third: the PL. I will formalize it and discuss the logical structure of its eventual falsification, addressing some possible criticisms of its utility. I will also characterize a specific form of it, the impossibilist law. Finally, I will conclude the chapter by showing in what sense PLs fundamentally differ from probabilistic laws, offering a unique and irreplaceable kind of prediction.

2.1 What Is a Natural Law? Three Answers

This section summarizes the most influential positions on the metaphysics of laws of nature to speculate: If a law of nature were these things described here, then what would be a PL? If the result becomes perfectly intelligible for all these positions, we may be on a good path. A caveat, first. We should remember that laws of nature are the real (in nature) laws that scientific laws aim to accurately represent. We must not confuse the two.

Here it is offered answers of what laws of nature could be summarizing the three most significant philosophical views on the topic: Hume's regularity account in its contemporary form (Lewis's best system account), laws as relations among universals, and the essentialist dispositionalism.

2.1.1 Best System Account

David Hume was a thoroughgoing empiricist. In being so, he held that we cannot know natural laws by any means beyond sense experience. For him, we cannot assume that the regularities we observe manifest a real invisible force that makes things behave in some ways and forbids them from behaving in others. Therefore, to him, laws of nature should be considered nothing more than descriptions of the regularities observed in the world through sensorial experience rather than physical or metaphysical necessities.

These regularities are based on repeated observations of events occurring consistently under similar conditions. If this is true, laws of nature are descriptive rather than prescriptive. They describe how objects or events have behaved in the past. That is the only reason we expect them to behave in the future under similar circumstances. However, we cannot extrapolate from finite past observations to infinite future cases. Therefore, Hume thinks they do not prescribe how objects or events must behave.

Here is the confusion in the Humean account. From our epistemic limitation of perceiving merely past regularities in our observations, some jump to the metaphysical thesis that there are no necessary or causal connections between regular events. There is a significant difference between claiming that one cannot rationally or empirically infer a causal link between the sunrise and the rotation of the Earth and asserting that no necessary or causal connection whatsoever governs the rise of the Sun. In other words, we may be unable to prove or even identify a genuine law of nature, but it does not mean they are not running the show.

Pedro Merlussi (2017) defends that "Humean" philosophers want to exclude any modal features from nature. They want a world without causes, necessities, and … possibilities. Merlussi thinks the ordinary Humean account of laws is naïve and intuitively problematic, as there are actual accidental regularities that are not laws. To illustrate this, he refers to Hans Reichenbach's example of the golden spheres: "All gold spheres are less than a mile in diameter" (Reichenbach, 1947, p. 368). Despite it being a genuine regular association, it does not seem that we can consider it a law of nature.

David Lewis assumed the task of giving Hume's account a criterion that distinguishes laws from accidental regularities. That is what we know nowadays as the "Best System Account" (Lewis, 1973b, 1983). According to Lewis, not every statement that summarizes a set of observations and their regularities is a law of nature. Laws of nature are only statements that belong to an ideal system of law-like propositions that offers the best overall balance between simplicity (fewest basic axioms and exceptions) and strength (broadest explanatory and predictive reach) in describing the regularities of nature.

In other words, the best system of laws is the simpler one, meaning that it should involve as few fundamental laws and exceptions as possible, as long as it is also the strongest, meaning it should cover the broadest range of phenomena possible. It is unclear, however, how we should choose, or weigh, between simplicity and strength when they point toward different candidate systems. The account itself does not specify how to strike the trade-off. This issue remains a permanent source of debate among sympathizers of this account.

Therefore, to this account, a true law of nature is not a result of a real state of forces in the world but a theorem that belongs to this ideal optimal set of minimal propositions that maximally explain our set of observations. We can also think of this set of laws, the best system, as the most elegant and comprehensive set of axioms from which we can logically derive all other statements about the universe's functioning. Formalized, a law in this position would appear precisely as a universal quantification of a material implication ($\forall x[P(x)\rightarrow Q(x)]$). We will analyze this logical form a little later.

This logical form represents a generalization about how certain conditions that entities or events must satisfy (P(x)) predictably lead to specific expected outcomes or behaviors (Q(x)), as captured by the predicates P(x) and Q(x). Such generalizations constitute laws within the "best system" if they contribute to achieving the optimal balance of simplicity and informativeness. Here, laws are contingent on the best possible scientific theory that describes the observed regularities of nature rather than being necessary natural forces.

2.1.2 Nomic Necessitation

David Armstrong was the original proponent of the view that a law of nature is a relation of nomic necessitation among universals. However, what does it mean?

A universal is a permanent property or attribute that something possesses (such as mass or negative charge), not a concrete contingent characteristic that something has at a particular moment (like the color of a chair). It is a property or attribute in itself, independent of any particular thing that merely instantiates these properties. For instance, a chair has mass; mass is one of its properties. But the concept of mass itself, and not the chair's mass, is a universal. A specific chair instantiates the universal of "mass."

Armstrong's key claim is that certain universals stand in a real necessitation relation, symbolized N(F, G), read "F-ness nomically entails G-ness." Certain universals necessitate the instantiation of other universals under certain conditions.

Take the case of "mass." "Mass," by its own nature, necessitates the attraction of "mass." In other words, the universal "mass" necessitates exerting a mutual gravitational pull. Every time "mass" is instantiated in the world, "gravitational attraction" also is. To Armstrong, we understand why regularities hold in nature by understanding the necessary connections between universals.

Let us put it even further. According to Armstrong (1989), laws of nature are contingent second-order relations among first-order universals. A law of nature, in this light, is also a universal, but a second-order universal. For example, the universals Mass (M) and Gravitational Attraction (G) stand in the relation N(M, G), meaning, "for all particulars x and y, if x and y instantiate Mass, they are nomically required to instantiate Gravitational-Attraction."

However, Armstrong is not an essentialist. This relation between mass and gravity is contingent; it could have been the case that it was not: "I have asserted that states of affairs such as N(F, G) are contingent. 'In a different possible world' it will not be the case that N(F, G)" (1983, p. 80). The universe could have been otherwise. The existence of this relation is not due to the essences of Mass or Gravitational Attraction; it is an additional fact about the actual world, and different N-relations could have obtained in other possible worlds.

That comes from his peculiar anti-Platonist approach of universals: "Universals are as much contingent beings as particulars, and they do not exist uninstantiated"

(Armstrong, 1983, p. 127). That is somewhat uneasy. The laws are true in virtue of a contingent second-order state of affairs. In the first order, ruling particulars, laws are necessary. If N(F, G) obtains, then every F is a G. But the law itself, that N(F, G) obtains, is itself contingent (it is not due to the nature of essences, and the world could have been otherwise).

If we should formalize what is a law in Armstrong's view, using the example of mass and gravitational attraction, it would be a function:

$$R(M, G)$$

where

- R represents the second-order relation (the law itself), indicating a necessary connection.
- M represents the first universal (a property or relation, like mass).
- G represents the second universal (another property or relation, like gravitational attraction).

This formula expresses that whenever a particular entity instantiates the first universal M, it will also necessarily instantiate the second universal G. What is fundamental here is the relation R, which indicates a necessary relation between M and G, in other words, a law of nature. For example, the law "All bodies with mass exert a gravitational attraction toward other bodies" could have M representing the universal "mass" and G representing the universal "gravitational attraction." The relation R would link these universals necessarily, expressing the law of nature.

2.1.3 Essentialist Dispositionalism

The recent rise of dispositionalism has given rise to its own version of the nature of laws of nature: essentialist dispositionalism. Alexander Bird is the best-known proponent of this theory of laws of nature, but he is not the first or the last. Cartwright (1989) and Mumford (1998, 2011) published influential versions of it.

Bird (2005) defends that properties are essential, inherent to the objects that possess them, and not accidental features. The dispositional properties of substances, or "capacities," are the tendencies for substances to behave in certain ways under certain conditions.

Dispositions, in that view, are reported by "if-then" relations, where the formula goes to "if certain conditions w obtain, then x behaves in way y." However, Bird (2005) argues that dispositions are irreducible causal powers whose manifestation is merely tracked (not constituted) by the conditional.

These dispositional properties, therefore, are grounded in substances' essential nature. Bird (2005) also refers to these dispositions as "causal powers," affirming that everyday objects possess inherent powers (in virtue of the dispositions of the substances that comprise them) that manifest in specific ways under certain conditions. These dispositions are not mere Humean regularities but are grounded in objects' essential nature. Laws obtain because of essential properties; they are nomologically necessary given those essences, not metaphysically unavoidable *tout court*.

Laws of Nature are, therefore, grounded in objects' dispositional (essential) properties. They describe the "if-then" regularities that result from the interaction of the dispositional properties of objects. Being so, laws of nature have an explanatory role because they have predictive power. Once they describe the dispositional properties of objects, they explain why the phenomena have occurred and predict which phenomena will occur (in certain circumstances).

Bird's (2005) dispositional essentialism rejects Armstrong's idea that the nomic relation is contingent. To instantiate a positive charge is to have the power to repel other positive charges, essentially. Therefore, the Coulomb law "charges of the same sign repel" holds in every possible world in which a positive charge exists. In this way, Bird eliminates Armstrong's anti-essentialist (contingentist) gap between properties and the laws they obey.

Despite the success of Bird's account, there are older versions of dispositionalism about laws. Such is the case with Nancy Cartwright's account of laws. Cartwright introduced the concept of causal powers as dispositions. However, she is not sure that there are any exception-free, fundamental laws.

Cartwright is one of the exponents of neomechanicism and one of the fiercest critics of the deductive-nomological model of explanation. She argues that explanations in science should be mechanistic rather than law-based. For her, natural laws do not apply in all contexts and circumstances. In her classic *How the Laws of Physics Lie* (1983), she argues that laws of nature are context-dependent. In other words, laws are not universally true but are contingent upon the specific conditions under which they operate.

In *The Dappled World* (1999), Cartwright defends that laws of nature are propositions about causal powers (dispositions). However, most of the time, what she calls "laws of nature" are the product of stable underlying and specific arrangements of substances or objects that she calls "mechanisms," and those mechanisms that produce natural phenomena. In other words, she treats capacities as local, mechanism-derived powers rather than essential dispositions. Cartwright's capacities are local and derived from mechanisms, not essential to substances in the Bird or Mumford sense.

Therefore, under Cartwright's view (and contrary to Bird's or Mumford's views), laws of nature are not explanatory in themselves nor fundamental features of the universe but are instead the result of underlying mechanisms that sustain regularities.

Cartwright's argument for the role of mechanisms in the production of the majority of what we nowadays call laws or generalizations is compelling. The problem is that she jumps from there to an anti-reductionist position, arguing that scientific explanations cannot be reduced to fundamental laws of nature. She believes that every level of scientific inquiry (quantum, molecular, biological, psychological, and social) requires distinct kinds of explanation based on the different mechanisms operating at each level.

The great appeal of the essentialist dispositionalist account of laws is that it provides a conceptual framework for understanding causality and regularities in nature, revealing the intimate relationship between causes and laws. Applied to the laws of nature, essentialist dispositionalism (Bird's and Mumford's) concludes that the laws are not external, imposed rules but are rather profoundly tied to the essential properties and dispositions of the entities involved. This position can be expressed formally as:

$$\forall x \, [D(x) \wedge C(x) \rightarrow M(x)]$$

where

- x is a variable standing for any entity.
- D(x) is a predicate expressing that entity x has certain dispositional properties (fragility).
- C(x) is the relevant condition obtains for x (being struck).
- M(x) is the manifestation that occurs (shattering).

That says: "for any entity x, if x has the disposition and the triggering conditions obtain, then x manifests that disposition." It is the relation that essentialist dispositionalists take laws to codify.

Thus, the laws of nature are physically necessary since they derive from the essential properties of things, and if these properties are physically necessary, the laws themselves are physically necessary. They could not be otherwise (given the physical nature of things). Moreover, the laws of nature are metaphysically explanatory, as they explain why things happen in the way they do, rather than merely describing how things happen. Laws provide a deep understanding of the causal structure of the universe.

Essentialist dispositionalism opposes the best system account of laws, which sees laws as contingent generalizations from empirical observations. Instead, essentialist dispositionalism sees laws as rooted in the very nature of things. Let us now see how laws work if each of these accounts were true.

2.2 Properties of Laws of Nature

The principal property of a law of nature is that it is always a logically contingent universal proposition; in virtue of its nomic (not merely logical) force, it supports counterfactuals. But what does it mean?

A logically contingent proposition is not a logical necessity (a tautology) nor a logical impossibility (a contradiction). Its truth cannot be logically determined. This kind of proposition is true in some possible worlds and false in others. They are also not true by definition; instead, they are true because of how the world is. In the same way, it could be conceivable that the world was different; thus, the laws of nature could have been different.

A law of nature also supports counterfactuals, that is, a conditional statement about what would be the case if a set of relevant facts in the world were different. This kind of reasoning (counterfactual reasoning) is crucial for science. It allows scientists to make predictions and inferences about situations that are not directly observed, because they are logically implied by the laws ruling them. For example, we can hypothesize that "If the mass of the Earth were twice what it is, the gravitational force between Earth and Moon would be double."

Because counterfactual reasoning enables us to imagine hypothetical scenarios, we can envision predictions and experiments, and scientists can test the implications of putative laws of nature under various conditions (Lewis, 1973b).

Let us look now for another set of properties. Merlussi (2017) asks, "Does the existence of laws imply that things could not happen other than the way they do, consistent with the laws staying the same?" (p. 22). As he argues, the answer to this question of whether laws allow for contingency depends on what the laws of nature are.

Merlussi and Cartwright (2018) draw our attention to four properties of laws of nature that should be determined so that we can know whether we can expect contingent facts in the world governed by them. They are:

- Extent: It is the range of the law, if it rules everything that happens under its cover. Thinking about L (the set of all actual laws of the universe), its potential extent refers to the entire set of happenings or events. So, an unlimited extent would cover the entire set of events in the universe; a limited extent would not cover at least some kind of domain of these events.
- Permissiveness: It is the variation that, given fully specified antecedent conditions, a law permits in the outcomes that fall under its prediction. A strict, sufficient law, given fully specified antecedent conditions, predicts one and only one event outcome; it is impermissive. A probability law, on the other hand, can accommodate more event outcomes in its predictions; it is a permissive law.
- Reliability: It is the level of certainty in which what the law predicts always happens. So, a law that always has its predictions fulfilled is totally reliable, but a law that sometimes has and sometimes has not is less reliable. If L were perfectly reliable, everything that it (plus the actual state of events) predicts would always happen.
- Potency: It is the power to realize in the world what the law predicts. In other words, are the things a law of nature predicts produced by it? If this answer is "yes," it is potent. Or are they merely happening according to its prediction? If the answer to the latter is "yes," it is impotent. On Humean best-system accounts, the law is descriptive (impotent). On nomic-necessitation or dispositionalist accounts, it is potent.

To exemplify these four properties, let us use the iconic (scientific, not natural) gravitational law, which we have already used here and will adhere to throughout this book for comparison. We can express this law by the equation $F = G (m_1 m_2/r^2)$, where F is the gravitational force between two objects, m_1 and m_2 are the masses of the two objects, r is the distance between the centers of the two objects, and G is the gravitational constant.[1]

Regarding the extent, gravitational law is unlimited, as for any two bodies of mass, it applies. In other words, there are no two masses that are not attracting each other to some extent. If it were any two bodies that would not attract each other, the gravity law would be limited to an extent.

Regarding permissiveness, Newton's law of gravity is impermissive, as it yields a single force value given fully specified masses and distance. In other words, they are not only attracting each other to some extent, but to the exact extent predicted by the law. If, supposedly, the gravitational constant would vary within some range, obeying some probabilistic distribution, then the resultant gravitational force would equally vary, making this imaginary law permissive.

The gravitational law is supposedly reliable since it is certain that what it predicts, once the antecedent conditions are satisfied, will always happen. We should not confuse ourselves here with events where the prediction of the law did not happen, as in the fall of a metallic object that stops falling right above a magnetic field. Those apparent violations stem from other forces, not from failures of the law; within its scope, the law is perfectly reliable. The gravitational force is being exerted here, as always; however, the object's trajectory is influenced by another force, namely, the magnetic force. To be unreliable, the law of gravity should have something chaotic about it when its predictions happen independently of other forces.

The gravitational law is potent depending on what the laws of nature are. If they are generalizations summarizing observations, as in Lewis's best system, the gravitational law is impotent. If they are an expression of the very nature of things, as in nomic theory or essentialist dispositionalism theory, the law of gravity is potent. In this case, it would be the law itself, as a manifestation of the nature of "mass," that is causing the effect of the force of attraction.

Now that these concepts are defined and exemplified, let us consider them applied to L, the complete and correct set of all the laws of nature. In that case, for the universe to be deterministic, it should be such that for any maximally specific state at a time t0, L fixes exactly one future (and past) history (van Inwagen, 1983). It requires that L has an unlimited extent, meaning that L covers the entire set of events of reality. That L be impermissive relative to complete antecedent conditions, and that L be perfectly reliable, meaning that what L says to happen always happens. Where one of these conditions fails—because L is limited, or permissive, or unreliable—objective contingency enters, and determinism collapses.

Let us read how Merlussi (2017) states the matter:

> Suppose that laws do not allow for contingency in the sense that (i) there are no happenings about which laws are silent, (ii) they are impermissive for agents and (iii) what they say to happen (with some relevant input) always happens. If (i), (ii) and (iii) hold water and the consequence argument is sound, then laws do not allow for contingency in the free will sense. (p. 23)

By (i), he means that an unlimited extent would cover the entire set of events in the universe, while a limited extent would not cover at least some kind of events ("happenings about which laws are silent"). In this last case, the universe would be indeterministic.

Merlussi and Cartwright (2018) identified the possibility of L (the complete and correct set of laws) being limited (in extent), permissive, or unreliable as the only way to an indeterminist universe. To them, these three possibilities are the only possible sources of contingency about what happens in the world. Ahead, I will reassess this thesis with the introduction of PLs.

By any of the three theories of the "law of nature," they are unlimited or universal in extent. In David Armstrong's nomic necessity and essentialist dispositionalism, it is obvious because these features are essential to substances. Therefore, they are always present and working when the substances are at work. Let us look closer at David Lewis's best system account (1973b, 1994) and Cartwright's dispositionalism, as they do not make things entirely clear.

Apparently, it would be possible for a bad generalization of a relation between types of events to be limited in extent. The problem with this is that if it were, it would not be a part of the best system of laws. The best system is an ideal set of the conjointly simpler and strongest laws. A generalization limited in extent would never be a part of it.

The question of extent in Cartwright's dispositionalism is trickier. It is a fact that the majority of what we call a "law of nature" nowadays is the product of the stable underlying and specific arrangement of substances or objects that we call "mechanisms." However, here, it is an epistemic problem. That is what we can discover in nature. It seems that Cartwright would not negate that, at a fundamental level, there are laws based on dispositional features of substances and that these fundamental laws are unlimited in extent.

Usually, when someone speculates about the possibility of L being limited in extent, they have in mind some specific domain of facts in the universe: the acts of a free agent. In that case, everything in the universe would be ruled by laws except the acts of consciousness. That would open space for free will. It is not an irrational belief, but rather than taking that route, I shall argue that PLs offer a stronger basis for genuine contingency.

None of the accounts of laws closes the matter regarding permissiveness. Some of the simplest and strongest generalizations could ideally be probabilistic (or possibilist) and, therefore, permissive. The nomic relation between two substances is the same. Some dispositions may be probabilistic (or possibilist). Therefore, there is nothing in the very nature of laws that makes them impermissive.

The problem of reliability is more intricate. Some generalizations within the best system would be unreliable. But not on the accounts of nomic necessity and essentialist dispositionalism. If laws are necessitated by universals or born from the very nature of substances, they will always manifest themselves in the same way. However, not on Cartwright's account if we are talking about the generalizations that emerge from the functioning of mechanisms, and not of the actual, dispositional laws of nature

Very roughly, this is the scenery of the quest for the nature and properties of laws. Now, we are about to consider the logical form and functioning of laws, preparing the ground to evaluate whether our definition and description of a PL possess the nature, properties, form, and functioning characteristic of a law. Finally, we will assess whether it possesses the characteristics of a scientific law.

2.3 How Do We Represent a Law of Nature?

Unlike the three metaphysical theories that compete about what a law *is*, the next three logical schemata each capture a facet of how scientists *use* laws in practice. While, at best, only one of the three theories of the nature of a law is true, here the two forms of laws that we use in everyday life in science will be shown, along with a third kind, that I defend we should use. I will call them sufficient, probabilistic, and—introducing a new category defended later—PLs.

However, before that, I prefer to ask if there is a minimal universal form that every law should hold to be a law. The best clue to start this discussion is to examine how laws are used to be represented mathematically or logically.

The mathematical form of a representation of a law of nature is the equation. Usually, we put the effect to be caused on the first side of the equation and the configuration of substances that cause it on the second side. It can be quite misleading. Let us start with our primary example here, the law of gravity. As we have already seen, this equation is usually represented as:[2] $F = G\,(m_1 m_2/r^2)$.

However, an equation does not represent a causal relation precisely because it is an … equation. It means that it is symmetric, and causal relations are asymmetric (A causes B does not imply that B causes A). For example, if we ask for the formula of the gravitational constant, our equation would be like this: $G = F.r^2/m_1 m_2$, and it is not intuitive to consider that the gravitational constant is caused by the product of gravitational force between two objects by its distance squared, divided by the product of its masses.

Let us examine another famous equation, Ohm's Law. It relates voltage (V), current (I), and resistance (R) in a single formula: $V=I\times R$, where V is measured in volts, current in amperes, and R in ohms. At first sight, the formula works intuitively (it always works, in reality), with the voltage being the result of a current flowing through a circuit multiplied by the resistance it finds along the way. But you can equally represent resistance as voltage divided by the current ($R=V/I$), and it is counterintuitive to think of resistance as being caused by voltage divided by the current.

Physical equations such as $F = G\,(m_1 m_2/r^2)$ or $V=I\times R$ express functional relations among variables. Algebraically they are symmetrical, yet practice supplies a causal orientation. Given (m_1, m_2, r) we treat the force as determined; given (I, R) we treat the voltage as determined. Rearranging the formula to solve for G or R changes the measurement target, not the underlying causal order (Woodward, 2003). Equations therefore need an interpretive layer to mark which variables are antecedent "conditions" and which are consequent "results."

That is why we examine logic for a more accurate representation of laws. Before, a caveat. Although the empirical test of a hypothesis does not require symbolic notation, giving a law an explicit formal shape is desirable for several reasons. First, a well-chosen formula forces the scientist to specify quantifiers, scope, and logical connectives that prose can leave vague. A clear logical structure is necessary for falsification, but formalization is not required. However, formalization proves that a hypothesis has a clear logical structure. Second, it makes hidden ceteris paribus clauses visible. Third, it flags type-versus-token slippage. Fourth, a formalized law can be fed into simulation software, combined algebraically with other laws, and checked mechanically for logical consequences. Formalization therefore improves communicability, comparability, and computability. In this book, given the history of obscurity surrounding the use of necessary conditions and "how-possibly explanations," I will employ formal expressions extensively, as a hammer to prove that we can use possibilist knowledge clearly. Formalism is a tool, a valuable tool, but not a prerequisite. A tool to facilitate the critique and improve the rigor, not to pretend that symbolism itself confers scientific status.

Continuing, a form that better captures the general characteristics of what we typically understand by a "law of nature" is a universal quantification of a material implication. This formula holds (with the same limitations) independent of any specific metaphysical framework about the nature of the laws. It can represent the universal extent, necessity, and conditional implication of a phenomenon given specific conditions altogether, and we have already seen its face:

$$\forall x[C(x) \rightarrow \mathrm{R}(x)]$$

where:

- ∀: universal quantification ("for all"), stating that the law applies to all instances within the specified domain.
- x: variable that can stand for any entity or set of conditions.
- C(x): predicate that describes specific conditions that x must satisfy.
- R(x): predicate that describes the outcomes that follow when the conditions C(x) are met.
- →: symbol for material implication, expressing a conditional: if C(x) is true, then R(x) follows.

We could read this formula as "for all x, if x satisfies condition C, then x has the outcome R." It states that this applies universally; every time the conditions specified by C(x) happens, the result R(x) follows. It is always important to remember that laws concern types (universals); tokens (particulars) merely realize them. When representing a law's application to a particular case, we should use the variable x, or replace it with a constant such as a, b, or c that names a specific individual.

A universal quantification of a material implication allows us to express laws rigorously as conditional statements that are universally true across all instances where the conditions are met. It captures the general idea that laws of nature dictate what happens under specific conditions. Moreover, it does it without committing to any metaphysical view of what a law of nature is (such as how these laws are grounded or why they hold). This logical structure is widely applicable and is compatible with various interpretations of what a law of nature is.

Sounds perfect. But it is not.

Material implication is not a perfect representation of metaphysical causation. Its limitations are significant, particularly in situations where antecedents that are completely irrelevant to the consequent can lead to true implications.

In classical logic, material implication can lead to situations where a true statement is derived from a completely irrelevant or even false premise. This can be confusing. The implication A→B is false only when A is true and B is false. This means that if A is false, A→B is always true, despite the truth value of B. For example, the sentence "if the moon is hollow, then I am a dragon" is considered true, even if the two component sentences are false. Even more concerning, the conditional sentence "if the moon is hollow, then voltage is current multiplied by resistance" is considered true, solely because the first sentence is considered false, and from something false, anything follows.

The two ways modern logic tries to deal with it are with counterfactual implication and relevance logic implication. Counterfactual implication considers what would be true in possible worlds most similar to the actual world where the antecedent is true (Lewis, 1973b). The problem is that it is not truth-functional because their truth depends on the relationship between "proximal possible worlds." You evaluate whether B is true in the closest possible world where A is true. If B consistently follows A in the nearest worlds where A holds, the counterfactual A → B is true (iff[3] in the closest possible worlds where A holds, B also holds).

In relevance logic, an implication A →r B (A relevantly implies B) requires not just that B be true when A is true but that there is a relevant connection between A and B.[4] This means A →r B aims to be true only when A contributes to B. The objective here is to avoid deriving true statements from false premises with no connection to the conclusion. The problem is that it is not truth-functional either, so it has no reason to be used to formalize laws.

Being conscious of these problems in capturing logically the functioning of causation is fundamental to putting formalization in science into its rightful place. But even with these limitations, we can do a little better.

Using temporal logic, primarily based on the work started by Amir Pnueli (1977), we can derive a formula that explicitly captures causal precedence. It quantifies over entities and conditions, supporting the temporal relationship to establish relevance between the condition and its result. With this resource, we can avoid most possible irrelevant relations (i.e., for all diachronic event laws). The result would be like this:

$$\forall x\, G(C(x) \rightarrow F(R(x)))$$

where

- G: Global Temporal Operator, which ensures that the relationship applies at all times.
- F: Finally Operator, which ensures that the result will occur sometime in the future after the condition.

The formula says that "for any entity x, globally (in any time), if C(x) holds t at a given time, then R(x) will follow later than t." It is an improvement on the traditional one, capturing the temporal precedence of causes over their effects.

However, this is not yet a perfect solution. Scientific laws such as Newton's Third Law (simultaneous action-reaction) or even Ohm's Law violate that template (Do the real laws of nature they intend to describe violate it as well?). A single temporal operator will not accommodate stateless, simultaneous, or bidirectional laws. However, it is serviceable for the causal cases we will examine here:

Sufficiency laws instantiate the template with a deterministic consequent.

Probabilistic laws relax the consequent to a fixed probability distribution.

Finally, PLs state what cannot happen when the relevant necessary condition is absent.

Thus, although no single formalism is perfect, the combination of functional equations, along with an appropriately strengthened conditional, provides a versatile toolkit for representing the diverse ways in which scientific laws guide explanation and prediction.

2.4 What Kinds of Scientific Laws? Three Answers

Therefore, the laws of nature can be represented, in its bare logical shape, as a universal quantification of a material implication, with temporal operators added. We could, for the sake of simplicity, keep the traditional form $\forall x[C(x)\rightarrow \mathrm{R}(x)]$ while tacitly assuming the temporal priority of C over R.

Scientific laws are our successful attempts to represent the laws of nature approximately. They can express them directly or indirectly. That means that a real law of nature can be different from the limited representation we are now able to formulate and test. Let us see what kinds of laws have the basic logical form and qualify as scientific.

2.4.1 Sufficiency Laws

What I call here "sufficiency law" (SL: that goes also for "strict law") is a traditional strict, deterministic law unlimited in extent, impermissive, and reliable. It represents a condition C sufficient to cause the result R and is conventionally understood always to produce a specific outcome under given conditions. An SL is typically characterized by its predictive accuracy and the assumption that if its initial conditions are met, its outcome will inevitably follow. We typically represent it as

$$\forall x[C(x)\rightarrow \mathrm{R}(x)]$$

noting that the condition in this formula necessarily happens first, then its results.

One of the most iconic and prototypical SLs is gravitational law. It states that any two objects in the universe always attract each other with a force directly proportional to the product of their masses and inversely proportional to the square of the distance between their centers. This means that the condition of having two masses m_1 and m_2 separated by distance r is sufficient for a determinate gravitational force to arise, with no exceptions.

In its logical form, gravity law would simply look like:

$$\forall m1,m2,r\ [Exist(m1,m2,r) \rightarrow GravityForce(F,m1,m2,r)]$$

Spelling out: "for all values m1, m2, and r, if m1, m2, and r exist such that m1 and m2 are masses and r is the distance between them, then there is a gravitational force F acting between them." From "GravityForce," we should understand the function that transforms the values of F: $F = G\ (m_1m_2/r^2)$.[5]

But if you take it wrongly, it becomes tricky.

Representing an ideal form is one thing; applying it in messy reality is another. The logical form we used here states that the result will follow whenever the conditions are met. However, as Cartwright (1983) has shown in her classic *How the Laws of Physics Lie*, such formalizations assume ideal conditions that do not exist in reality; invariably, they overlook complexities such as other concurrent forces that alter, at least minimally, outcomes in real scenarios.

Cartwright contends even further that many (considered) laws of physics are not universally true, as they do not describe reality but are no more than models, simplified idealizations for the sake of calculation. They are not universally true because they are ceteris paribus laws, which require the clause that all other influencing factors are held constant. These conditions are never met in the real world, making the laws not universal.

Cartwright's critique targets idealized scientific laws, not the possibility that real laws of nature exist. Laws of physics generally describe how things would behave in highly idealized and controlled conditions, like those achieved in the laboratory, rather than in the uncontrolled natural world. Cartwright argues that what is real in the world are not abstract laws but objects' causal powers and capacities (the "dispositions"). These powers and capacities give rise to the phenomena we observe rather than the abstract general laws of science.

Let us return to Cartwright's ideas when analyzing the problems in the deductive nomological model of explanation and mechanistic explanation. For now, it is important to make some final remarks on strict laws. Cartwright's work is a cornerstone of the end of innocence about a logical conception of laws of nature. However, we should not confuse two things. First, we should not confuse science's messy and imperfect laws with the real laws of nature. In other words, we should not confuse our epistemic limitation with a supposedly messy metaphysics. Second, we should not take laws of nature, as Cartwright has taught us, as logical entities that determine logical consequences in the world. The laws of science (not the laws of nature) are logical or mathematical representations of real causal forces (ruled by laws of nature). If this is the case, we can expect that the real force represented by the equation of the law of gravity is always exerted, even when, because of the intercourse of other forces (friction, magnetism, propulsion), the result of the formula is not obtained. In short, a law of nature is a causal entity that can be captured, albeit imperfectly, by a logical schema.

2.4.2 Probabilistic Laws

A probabilistic law states that under certain conditions a particular outcome occurs with a specific probability rather than inevitably. One of the differences between a probabilistic law and an SL is that the former is not deterministic but permissive. Probabilistic laws are very common in genetics and quantum mechanics and totally dominant in disciplines that deal with extremely complex environments, such as psychology and epidemiology.

From the beginning, it is essential to distinguish between the epistemic and the metaphysical reasons for using probabilistic laws. We can use them to raise our prediction capacity in systems where outcomes vary under apparently identical conditions. And we think they seem identical, probably because of our limited knowledge of these systems. That is the primary reason for their use, an epistemic one. However, there may be nature's domains where randomness is inherent to it, and the corresponding scientific probabilistic law does not reflect only our limited scientific knowledge of the systems or mechanisms but the actual law of nature, the very structure of reality, like in quantum mechanics. This, if true, would be our metaphysical reasons.

A probabilistic law can be expressed in logical-probabilistic form:[6]

$$\forall x\ [C(x) \rightarrow P(\text{Outcome}|C(x))=p]$$

where

P(Outcome|C(x)) represents the conditional-probability function.

p is the probability that an outcome occurs, given the condition C.

We can read that as "for every instance x where the condition C(x) is satisfied, the probability of a particular 'Outcome' given C is exactly p."

For example, a probabilistic law in genetics might state that the probability that a child inherits a certain trait (like the color of the eyes) from heterozygous parents follows a probability distribution:

$$\forall x[\text{Parents}(x,\text{Heterozygous})\rightarrow P(\text{Child}(x,\text{BlueEyes})|\text{Parents}(x,\text{Heterozygous}))=0.25]$$

We can read this as "for all children x of heterozygous parents, the probability that the child has blue eyes, given that the parents are heterozygous, is 25%."

Although rooted in genetics, it is quite unlikely that, at the fundamental level, the process of mixing DNAs is physically random. Probably, when the two DNAs mix into the zygote, subtle variations in the zygote's environment (the uterus) may sufficiently determine the result. But this remains speculative, we do not have, presently, the knowledge to answer the question.

This is the major criticism about probabilistic laws: they would be only "step laws." They would be the result of a limitation in fields where the real laws of nature (deterministic) are not at hand. Determinists argue that the world is deterministic at the fundamental level. Therefore, probabilistic laws would be merely reflections of our ignorance of these actual fundamental laws. As we only recognize some contributory conditions influencing the result of a situation, our law becomes probabilistic because it reflects only the known conditions. Is it really so? I will come back to it in the Psychology chapter.

This nature of probabilistic laws also raises criticisms about their explanatory power, as we will see in the next chapter. Critics argue that probabilistic laws describe how often phenomena occur but not why they occur in individual cases. In psychology, for example, they would be unable to explain individual behavior. Probabilistic laws would only be capable of describing how phenomena occur rather than explaining why they occur.

Moreover, probabilistic laws do not guarantee that conditions lead to outcomes; they only increase the likelihood of outcomes. This complicates interpreting causes, especially in complex systems with multiple causal factors. Probabilistic laws usually rely on statistical data from these complex systems. Therefore, they are susceptible to all difficulties typical of statistical analysis, such as sampling errors, bias, and data quality. This compromises the reliability of these laws. Again, we will return to it when analyzing comparative ceteris paribus laws.

2.4.3 Possibilist Laws

We are finally entering the open sea. A PL, instead of stating a sufficient condition or a probability function, states a possibility condition, in other words, a necessary condition for the occurrence of a phenomenon. The traditional caveat to the possibility of such a scientific law is that predictions of possibilities are not falsifiable because whether the "predicted" event occurs or not, the "prediction" would be satisfied. It would be something like an astrological prediction. Well, this is not quite like this.

The presence of PLs in this mini subitem is just to mark their rightful place among the other scientific laws. However, as I am construing this entire chapter to understand them better, I am opening a new item specifically to analyze them.

2.5 How Does a Possibilist Law Work?

PLs and comparative ceteris paribus laws (Rupert, 2008) may be the only kind of laws possible for the human sciences. Instead of stating a sufficient condition ("if x, then y happens") or a probabilistic condition ("if x, then y has p of probability to happen"), they state a necessary condition, taking the negative form: "if not x, then y cannot happen" (Castañon, 2021a). That is how real prediction and falsification of possibilities become possible.

2.5.1 The Mess with Necessary Conditions

Necessary conditions, therefore, are the core of possibilist thinking. However, as noted by Goertz and Starr (2003) in the first book dedicated to the methodology of studying necessary conditions, this concept "seems deceptively simple" (p. 1), and it is easy to commit mistakes in theory, logic, and methodology when dealing with it. Because logically distinct ideas share the label "necessary condition," scientists may doubt they can underwrite laws. Even Goertz and Starr speak only of "necessary condition hypothesis." At the end of this item, I will retain one strict sense of "necessary condition" only for law-building.

Hart and Honoré (1985) provided one of the first accounts of the confusion surrounding the use of this term, particularly in legal contexts. To them, in ordinary language, we use this term to refer to at least three different things.

First, following them, we can call a condition "necessary" simply because it is one member of a set of jointly sufficient conditions to produce some effect. In this case, "it is necessary because it is required to complete this set" (Hart & Honoré, 1985, p. 112). To illustrate this and other senses of "necessary condition," let us stick with Mackie's (1980) classic example of a house blaze. A blaze in a house was produced by an overloaded electrical circuit (condition), dry wood in the attic (condition), and the absence of a fire-sprinkler system (condition). This set of conditions was sufficient for the fire, though none alone was. However, in the sense of "necessary" used here, all three conditions fit. Each one was necessary for the set of conditions to become sufficient for the fire to occur. Take out one of them, and the blaze (effect) would not happen. This is a weak sense of "necessary condition," as an overloaded electrical circuit could have been substituted by a match or a lighter, dry wood in the attic by thousands of books, and the absence of a fire-sprinkler system by … the absence of almost anything. Hart and Honoré do not spell out why such factors are only "episodically" (in my term) necessary for E. I label some of these factors "permissive conditions," a distinct logical and causal type of condition to be identified in science.

The second sense in which we use this term in ordinary language, according to Hart and Honoré (1985), is stronger, and that is the only sense in which I am using the term "necessary condition" in this book. A "necessary condition" for E would be a "member of every set of sufficient conditions of a given type of occurrence" (p. 112). An example from Mackie's story is oxygen: no fire can ever occur without it. We could say that a necessary condition for E must be present in any possible world, with the same natural laws as ours, in which E occurs. PLs are made of these conditions because they apply unconditionally, in all particular instances of some universal. Another way to say it is that laws concern types; tokens merely realize them. In this sense, dry wood is not a necessary condition for a blaze but exposed inflammable material. An overloaded circuit is not a necessary condition for a blaze (any blaze), but an ignition source. In that particular case, the dry wood instantiated the necessary condition for a blaze: exposed inflammable material. It was a *permissive condition* to E: it was sufficient to bring about that necessary condition. However, in another particular case, the inflammable material of dry wood could be wrapped in an isolating material and could not have been a permissive condition to E (a sufficient condition to N). We will elaborate on this definition and differentiation further in Chapter 3.

The third sense would be *sine qua non* conditions. Of course, the first and second sense of necessary conditions are also *sine qua non*. However, to Hart and Honoré (1985), there are *sine qua non* conditions that are only necessary for the event on some occasion. Suppose that in the house blaze, there is not only dry wood in the attic but also thousands of old books. In this case, thinks Hart and Honoré (1985), the dry wood would still be (episodically) necessary in the first sense (it is a member of the set of conjointly sufficient condition to the blaze), but it was not a *sine qua non* condition to it (the books could have fueled the blaze instead). I think this position is quite misleading. In this case, it cannot be said that the blaze would not happen given the short-circuit. However, they give other kinds of examples of "irrelevant *sine qua non* conditions" that would illustrate a third kind of use of the term "necessary condition." They are the "analytic connection" (p. 114) and the "incidental connection" (p. 115).

An analytic *sine qua non* is a logical or linguistic necessary connection of something with an event, which, however, has no causal effect on it. Imagine the event of someone becoming a widow because of the murder of her husband, James. It is possible to say that a necessary condition for the woman to become a widow was the fact that she got married. If she had never married, she would not have been a widow now. That is a necessary condition between the words used to express a fact, not to a fact. If we talk about the fact in other terms, the sentence would become unjustifiable, like in "If she had never married, James would not have been murdered."

A more bizarre *sine qua non* condition is a causally irrelevant incidental connection to an actual necessary condition. In any causal concrete connection, we can also add more detail in the description of the process that led to the effect. Suppose that James was killed by a man who shot him. So, the shot was a *sine qua non* condition to his death. However, the killer did it wearing a suit and a black glove, with the left hand, and using a 38-millimeter caliber, not 45. So, these unique features of the crime identify it uniquely, in the sense that someone can say, in a court, for example, that "if the man wearing a suit and a glove in his left hand had not fired a 38-millimeter gun, James would not have been killed." As Hart and Honoré (1985, p. 116) say, "to assert this is to say no more than this particular act of shooting, uniquely identified in this way, caused the death." Their third use hinges on uniquely specifying an act (glove-wearing shooter) rather than adding a causal factor.

Summarizing, Hart and Honoré's diagnosis needs refinement. I will work here with three other categories.

First, a "necessary condition" N for a kind of phenomenon Z is something such that, if N fails, Z cannot occur—across all possible worlds that share our laws (an ignition source N for a blaze Z). This concept is also captured in the second use of Hart and Honoré (1985). Here, against them and Goertz and Starr (2003), I treat background necessary conditions (as oxygen) as genuine causes to be considered, especially in possibilist explanations. The motive will become clear in Chapter 3.

Second, a permissive condition P for event E is a factor whose presence, in a given context, brings about or activates some necessary condition N for E (a kind of Z) in some particular context (a short-circuit P for an ignition source N). This definition encompasses most cases from the first use of Hart and Honoré, as well as some of the third.

Finally, the third type would be irrelevant *sine qua non* conditions, which should be excluded from causal explanations and distinguished from permissive and necessary conditions. Note that I do not include background necessary conditions in this category. Analytic *sine qua non* and incidental *sine qua non*, however, are examples of it.

2.5.2 The Form of a Possibilist Law

This section builds upon my earlier formulation in Castañon (2021a). The formalization of a PL could be like this:

$$\forall x\ (\neg Nx \rightarrow \neg\Diamond Rx)$$

where

- $\forall$ is universal quantification ("for all").
- *x* is a variable that can stand for any entity.
- N(x) is a predicate that describes a necessary condition to R that x must satisfy.
- R(x) is a predicate that describes what possibly follows when condition N(x) is met.
- $\rightarrow$ is the symbol for material implication, expressing that if $\neg$N(x) is true, then $\neg\Diamond$R(x) follows.
- $\neg$ is the negation operator.
- $\Diamond$ is the modal operator for possibility.

The result, written in regular English, would be "For any entity x, if x does not satisfy the necessary condition N, then it is not possible for x to achieve the outcome R."

Formalizing PL with the simple negation of the result ($\neg$Rx) would be possible without the modal operator of possibility ($\Diamond$). It would be: $\forall x\ (\neg Nx \rightarrow \neg Rx)$. This makes no difference when considering its predictions and falsifications, but it would be a loss when explaining the possibility of a seemingly epistemically impossible event, as we will see in the next chapter. Therefore, for the sake of the possibilist explanation, formalization with the modal operator is needed.

2.5.3 The Falsification of a Possibilist Law

When addressing the issue of falsifiability for PLs, it may initially seem counterintuitive to predict possibilities since the presence of x might seem to cover any result. However, what a PL actually predicts is the impossibility of an outcome without its necessary condition. For example (Castañon, 2021a), asserting that "only in the presence of oxygen can combustion occur" means that combustion cannot occur without oxygen:

(i)	$\forall x\,(\neg Nx \rightarrow \neg\Diamond Rx)$	(possibilist law)
(ii)	$\neg Nx$	(there is no necessary condition N satisfied)
(iii)	$\therefore \neg\Diamond Rx$	(therefore, there is no possibility of result R)

The PL is a scientific law because the eventual occurrence of an effect without the necessary condition falsifies the law. A PL, in practice, does not predict that x makes y possible (which does not predict anything falsifiable: "oxygen allows combustion"). The structure of falsifying a PL is as follows (Castañon, 2021):

a) PL1: "If there is no N (oxygen), there is no R (combustion)."
b) It is the case that we have R (combustion) without N (in a chamber without oxygen).
c) Therefore, PL1 is falsified.

We should never forget that the *logical* ease of the falsification of a PL does not mean *practical* ease. The problem of the "empirical base" of falsification (Popper, 2002, p. 74) has been a topic of discussion since Popper's original work in 1934. There is a usual confusion between logical refutation, which is certain, and empirical refutation, which is as conjectural as any other scientific hypothesis. Logical refutation is absolute: from the falsity of a deductive consequence of a set of premises, it necessarily follows that at least one of its premises is false. However, with methodological falsification, the process is much more complex and sophisticated, and—like everything involving the empirical—it is also conjectural. In other words, for Popper, an assertion that falsifies a theory is itself fallible, but intersubjectively testable.

In or out of the laboratory, falsifying a PL would never be as simple as recording a single countercase. One must also establish, to an agreed tolerance, that the necessary condition N was genuinely absent. That requires well-calibrated instruments, independent replication, and auxiliary hypotheses about background interference or hidden variables, as always.

Scientists, therefore, should report error bars (or detection thresholds) for N, run control experiments, and, when feasible, seek multiple convergent indicators of N's absence. Only when those checks survive peer scrutiny does the community treat the observation of E as a *bona fide* refutation of the PL.

Moreover, there is always the problem raised by Pierre Duhem (1906/1954) about the fact that any putative falsifier tests the whole bundle (PL + auxiliaries + measurement model) of theories presupposed in an experiment. In other words, when a hypothesis is falsified, the error could reside anywhere in the long chain of theories and auxiliary hypotheses used to construct and test it. Later, Quine (1951) sharpened the description of the problem, pointing out that every statement—even logic and mathematics—belongs to a holistic "web of belief" that can be retained or abandoned in response to observations.

However, this is a manageable problem for the falsifiable and approximative image of science advocated by Popper. To deal with it, we should have a way to conjecturally identify, through rational choices, the false hypothesis within this tangled web of auxiliary assumptions. And there is a way.

As Popper (1979) addressed in Objective Knowledge, and Andersson (1994) elaborated further, when we conjecturally single out a hypothesis as suspect, it must be subjected to independent tests. That means tests that do not presuppose the hypothesis being evaluated and instead rely on theories currently deemed unproblematic and well-established (with no conditional falsifications). In other words, as argued by Andersson (1994), although every theory is tested through experiments and the theory-laden observations derived from them, we can avoid circularity only using tests that (while fallible and theory-dependent) do not rely on the problematic theories under scrutiny.

For example, if one looks through a telescope and sees moons around Jupiter, an ardent Aristotelian could have argued that the telescope somehow creates those images due to a flaw in the lens. However, since the theory at stake is the existence (or not) of the moons (with Copernican theory in the background), corroborating evidence from the telescope can be tested independently of the astronomical theories in question by pointing the instrument (as Galileo did) at a distant but well-known church tower on Earth. Here, it is not the contested theory being presupposed but another one: optics.

2.5.4 Impossibilist Laws

There is a subclass of PLs that I call "impossibilist laws." While PLs express necessary conditions for the possibility of an event, impossibilist laws denote absolute impossibilities in nature for any physical entity with a specific predicate. This means that no necessary condition can make the outcome possible for any x, in any situation. They can be formulated as:

$$\forall x\,(Px \rightarrow \neg\Diamond Rx)$$

where

- *x* is a variable that can stand for any entity.
- P(x) is a predicate that describes a condition that makes R(x) impossible for any entity *x* that satisfies it.
- R(x) is a predicate that describes the outcomes that cannot follow when the condition P(x) is met.
- $\Diamond$ is the modal operator for Possibility.

In plain English, this reads: "For any entity x, if x satisfies the condition P, then it is not possible for x to achieve the outcome R."

Some examples before more enlightenment. Reaching absolute zero temperature is impossible in physics, no matter what physical system we consider. We can express that as: "For all physical systems x, it is not possible for x to reach absolute zero temperature."

Why cannot we formalize this as $\forall x\,(x \rightarrow \neg\Diamond Rx)$? Because this formulation would apply to any x, including nonphysical entities like "circles." This would be fundamentally wrong, as logical entities have no temperature (or could we consider their temperature to be zero because they don't have any?). There are impossible laws in geometry; for example, "for any x, it is impossible to be a round square." However, when talking about empirical science, we restrict our enunciations to physical entities. Therefore, using predicates (as "physical entity") is essential, especially because regional impossibilist laws apply to specific classes of phenomena.

For example, we can say that "for all neurons *x* in the adult human prefrontal cortex, it is not possible for *x* to undergo neurogenesis" (to regenerate). In such cases, the predicate is paramount: regeneration is considered impossible only for adult neurons in the prefrontal cortex. Of course, falsifying this law would be even easier than falsifying a PL: it suffices to point to one case of neurogenesis in an adult prefrontal cortex.[7]

It is interesting to note that a scientific discovery in such cases would transform the impossibilist law into a PL by identifying a necessary condition for neurogenesis in the adult prefrontal cortex. Assume that a hypothetical treatment or genetic modification (T) that enables neurogenesis is discovered. The former impossibilist law would have been transformed into a PL, which says that "For any prefrontal x, if x does not undergo T, then neurogenesis R is not possible" ($\forall x\ (\neg Tx \rightarrow \neg\Diamond Rx)$).

To evaluate an alternative formalization of impossibilist laws, consider why we should not use $\forall x\ (Px \rightarrow \neg R(x))$ without the modal operator. While this formula expresses the same events, it does not explicitly involve the modal aspect of possibility. It simply states that R(x) does not occur when x occurs. Including the modal operator $\Diamond$ in the formalization is important for possibilist explanations, as it captures the nuanced idea that certain outcomes are not just false, but fundamentally epistemically impossible.

Another problem with impossibilist laws is the problem of triviality. Evidently, the majority of impossibilist laws are trivial, as "if x is a human, then x being planet Venus is impossible." However, similarly with PLs, what matters to science is the nontrivial impossibilist laws, as "it is impossible for any object with mass to travel faster than the speed of light."

2.5.5 Is a Possibilist Law Really a Law?

Essentially, a PL is just a specification of a necessary condition for some type of event. That does not seem all that new. What is new here is that such specifications are considered scientific laws, proper representations of laws of nature. And we should consider these specifications of necessary conditions as laws because they are logically contingent universal propositions and, therefore, support counterfactuals. Moreover, they are not just laws, but strict laws, probably the only type of actual strict (scientific) laws in science, and possibly, of strict laws of nature in the universe.

Let me fulfill my promise and justify this last affirmation. Traditionally, laws (SLs) have been considered a relationship between a set of descriptions of a set of conditions at the situation background (B), plus the description of one type of event that is a particular condition (C) specified in the antecedent, with the description of another type of event that is the result (R). R is the result of the particular condition *plus* the set of conditions in the background (C ∧ B). This relationship is such that it guarantees that if a specific background B and an event C occur, event R must occur. Therefore, what a strict law really means is (B ∧ C) → R.

However, as we know very well nowadays, establishing such an actual strict SL in science is practically unrealizable (Cartwright, 1983). To establish sufficient conditions for R, we would have to include and discriminate the entire set of conditions that compose the background conditions B of the law; in other words, all necessary conditions whose presence or absence could prevent the occurrence of R.

A PL does not have this problem because the only thing it predicts, in practice, is the impossibility of R in the absence of one and only one actual necessary condition N and transforming its negation (¬N) into the entire antecedent, as well as the negation of the event R (¬R) into the consequent. No background clauses. And it is a strict law because the negation of the necessary condition (¬N) sufficiently implies the negation of the possibility of the event result (¬R). It is a real and practical law, predictive, falsifiable, and ready to be used by anyone who wishes to prevent, not induce some event.

This implies something else, amazingly counterintuitive. Even being a strict law, as the result (or effect) of a PL is the absence of something, it does not determine what will happen, only that something will not happen. Let us imagine a scenario where L (the entire set of actual laws of the universe) were entirely composed by PLs. In this universe, everything would be entirely ruled by laws and entirely indeterminist. L, in this case, would admit an unlimited set of different outcomes, remaining silent about their probabilities.

This characterization presents us with a unique kind of law: one that is indeterminist, as it cannot determine how the world will be, only what the world will not be. It is a law unlimited in extent, supposedly ruling the entire domain of the kind of event-result. It is impermissive, not allowing variation. Once N is negated, R is always negated. Finally, it is a reliable law, as its prediction is expected to hold 100 percent of the time.

2.5.6 But … What Is Possibilist Law for?

How often do we come across PLs in the wild? Since we are used to thinking of a de facto necessary condition as the complete set of conditions that can individually instantiate an actual necessary condition, this may seem very difficult to find. However, this way of thinking is wrong, and finding PLs is extremely easy. We are surrounded by countless easily identifiable PLs every minute of our lives, everywhere. That may be why we think so little about them.

Let us analyze the problem of looking for necessary conditions considering Mackie's model (1980). A necessary and sufficient condition for an E event (lighting a candle) must be a conditional in which the antecedent is a complex disjunction of conjunctions (light a match and touch your head to the candle wick; or place a lens in the sun and focus your rays on the candle wick; or fire a gun on the candle wick; or … etc.), each conjunction representing a condition that is sufficient but not necessary for E. This is the case because there are generally several ways to cause an event (lighting a match and touching it to the wick of a candle is enough to light it, but it is not necessary because I can also do this using a lens, a gun, a torch, etc.). These various forms make it impossible to have a "general law sufficient to light a candle" but not a PL to do so.

A necessary condition N for E can be found in two ways. First, it may be a hidden (background) condition in the "causal field" (the presence of oxygen to combustion). Mackie (1980) calls the "causal field" the entire field of conditions that are naturally present in the context of E, which must remain unchanged (that the world exists, its typical environment, oxygen, etc.) for it to occur. These basic background premises (conditions) are, in a strict sense, part of the cause of E, but they remain disregarded most of the time by us for pragmatic reasons.

Second, we can find it looking for some component that is present in all members of the complex "disjunction of conjunctions" that Mackie said to be a necessary and sufficient condition for an R event (light a match and touch your head to the candle wick; or place a lens in the sun and focus your rays on the candle wick; or fire a gun on the candle wick; or … etc.). In the example of lighting a candle, it would be heating the candle wick to a temperature sufficient for combustion. All the members of the disjunction sufficiently heat the wick, it is a non-redundant part (its function is not realized by other condition) of all conjunctions (light a match and touch your head to the candle wick; or place a lens in the sun and focus your rays on the candle wick; or fire a gun on the candle wick; or … etc.) that are minimum sufficient conditions for E.

Therefore, PLs are abundant in nature. What is not so easy to find is a PL that is not trivial and presents different insights and uses. While it may seem trivial in the twenty-first century to state that if there is no oxygen, there is no flame, this simple PL still holds practical value. It can be effectively used to prevent combustion.

An obvious objection to predictions from an experimental test of a PL would be that since event E cannot be guaranteed by the occurrence of necessary condition N, we cannot cause it in the laboratory simply because the PL does not say how we imagine instantiating it. In other words, how do we produce E if we do not have sufficient conditions to cause it?

Well, we do not need to do that. I should point out two interesting features of the PL here. First, any E reliable observation in the absence of N falsifies a PL without needing an experiment. PLs are not ceteris paribus laws. They could be an excellent tool for social sciences, as the latter cannot reproduce the social effects they investigate. PLs offer predictions that can be falsified without experimentation with a single reliable observation.

Consider Marx's famous prediction that no social order disappears until all its productive forces have been developed. If we interpret this generalization as a PL, just one instance of revolution in an agricultural country in the industrial era would be enough to falsify it. That is precisely what has happened. Goertz (2012) believes that there are many of what he calls "descriptive causal generalizations" in the social sciences and that the most common of these is the description of the conditions for an event. What discourages many social scientists from presenting such strong generalizations as laws, he thinks, is a strategy of protecting their findings from a counterexample. In another work (Goertz, 2003), he compiled a list of 150 hypotheses of permitted conditions in history, sociology, political science, and economics, several of which could be considered candidates for PLs. For example, we have the generalization of "democratic peace," a thesis that can be stated in a possibilist way: "If neither of two states is authoritarian, the war between them is impossible."

The second interesting feature of the PL to be considered is that, in fact ... it could be manipulated. However, this only happens in one discipline. Psychology is in a unique position among the sciences to leverage PLs practically. We can conduct psychological experiments to test PLs. By removing the antecedent necessary conditions and asking subjects to demonstrate behavior E, we can offer a reward as an incentive. If they fail to execute E, we have substantial grounds to consider the law supported. For instance, speculating that a preserved temporal lobe is a necessary (not sufficient) condition for the emission of verbal behavior, we can test this generalization by subjecting people to transcranial magnetic stimulation (TMS) inhibiting the area and offering experimental subjects a reward for speaking. In other words, in psychology, we have the unique ability to attempt to achieve E while suppressing an N because the samples in it are conscious agents. It made this discipline making it an ideal field for testing PLs. Let us come back to this point in the Psychology chapter.

2.5.7 The Difference between Possibilist and Probabilistic Laws

Let us investigate possibilist and probabilistic laws, where neither is a foolproof predictor of specific events. While a PL predicts a nonevent, a probabilistic law only informs us of the chances of some occurrence. At first glance, probabilistic laws may appear to hold more predictive power than PLs, and it is possible that PLs can be derived from them. However, as we will see, this is very unusual.

Thinking in mathematical terms, the probability calculus already includes the modal extremes: $P = 0$ corresponds to impossibility, and $P = 1$ to certainty. However, that does not mean they are ordinary cases of probability. They are the PLs and the SLs, and, in being limit cases, can be expressed and tested without the characteristic uncertainty of probability calculus.

A bare probability statement $P(\neg Y|\neg X) = 1$ tells us that the probability of Y not occurring given that X will not occur is one, which is equivalent to my formulation of a PL. Another way to express a PL in probabilistic terms is $P(Y|\neg X) = 0$.

2.5.7.1 Methodological Differences

However, there is no complete falsification in probability. A single counterinstance cannot confirm an actual zero or one; statistically, no finite sample can. In contrast, a PL is tested by manipulating N: remove N and check whether R becomes impossible. This experimental handle is absent from a purely probabilistic framework.

My point, therefore, is both epistemic and methodological. When we build a probabilistic model, we choose a sample space and a measure. Sometimes that measure assigns probability 0 to an outcome only because of a convention, not because it is ruled out by nomic necessity. Suppose one throws a dart at a square target, modeled as a Cartesian system of coordinates. The probability of the dart hitting any exact point is 0 (compared to the total area, the point occupies zero extension). However, the dart can still hit exactly point (7.0, 7.0).

2.5.7.2 Possibilist Law Says What to Manipulate

Another example. Let us take the example of phonological working-memory tasks (Gathercole, Pickering, Ambridge & Wearing, 2004), where free capacity ($\neg N$) is not available; complex reasoning (R) is impossible. A probabilistic summary might report $P(R|\neg N) \approx 0.00$ in a sample. However, only the PL "Within a saturated phonological loop, complex reasoning cannot occur" reveals what to manipulate and why the effect persists across labs. A PL picks up a manipulable necessary condition. Removing N is a test; adding N is a design constraint. A probabilistic model seldom tells which component to tweak.

Probability zero may stem from modeling choices and still leaves room for "zero-probability events." A PL ties the event to a necessary condition, making its nonoccurrence a matter of logical consequence, not of negligible measure.

Many domains involve structural or logical constraints that probability alone cannot capture. A bacterial cell without a chromosome has a probability of 0 of dividing, not because of stochastic improbability but because the mechanism lacks its necessary condition. Treating this as a possibilist relation clarifies both the impossibility and the relevant biological mechanism, rather than mathematically complicating something logically simpler.

2.5.7.3 No Zero or One Probability Can Come from Step Laws

Understanding the type of probabilistic law we are dealing with and the kind of prediction we seek is crucial. An SL can predict the occurrence of an event, while a probabilistic law can estimate the probability of its occurrence. On the other hand, a PL can strictly predict the nonoccurrence of an event, an objective that a probabilistic law cannot always achieve. These distinctions have real-world implications that we will explore further.

For example, let us remember the example of probabilistic law in genetics that we used. The probability of a child inheriting a certain genetic trait (like eye color) from heterozygous parents follows a specific probability distribution: $\forall x(\text{Parent}(x,\text{Heterozyg}) \rightarrow P(\text{Child}(x,\text{BlueEyes})|\ \text{Parent}(x,\text{Heterozyg}))=0.25)$

Remembering this genetic simplification, a heterozygous individual is one carrying two different alleles of a particular gene. In this case, one is recessive (blue eyes: b), and the other is dominant (black eyes: B). When two alleles (one of the mother and one of the father) combine in the child's gene determinant of eye color, the dominant allele will always prevail if they are in the gene (BB, Bb, bB), so the only chance of a blue-eyed child is two recessive alleles combined (bb).

That is why this statement asserts that for all children x of heterozygous parents, the probability that the child has blue eyes is 25 percent. As this probabilistic law is objective and rooted in an apparently random natural process, we can, if we want, extract a prediction of the impossibility of a child born with blue eyes. If one of the parents is homozygous with black eyes, the probability of a child with blue eyes is zero. Similarly, if the two parents are homozygous with blue eyes, the probability of a child with black eyes is zero. Of course, in medical practice a residual risk remains due to undetected rare alleles or laboratory error. Still, the logic of the PL is categorical once the absence of a known mutation is taken as established.

We could transform those two probabilistic laws into PLs by stating that the necessary condition for a child to be born with blue eyes is that both parents have at least one recessive blue allele. If this necessary condition is not satisfied, then the child cannot have blue eyes. The second PL would be that if neither of the two parents has at least one dominant allele, then it is impossible for the child to be born with non-blue eyes.

Nevertheless, imagine that we want to know if it is possible for a single woman to have a child with blue eyes one day. We cannot set the probability of this happening since we do not know if she can be pregnant, if she is going to have sexual relations with a fertile man or be inseminated, or, finally, what the genetic constitution of the putative father is. However, if we know that she has no recessive blue allele, according to the law of possibility that "if a woman has no recessive blue allele, then she will have no blue-eyed children," we can make a strict prediction that she will never give birth to a child with blue eyes.

Moreover, we are working on a very specific case here where probability appears to be objective, and there are very few examples of it in science. Most probabilistic laws are derived from complex environments, reflecting the limited understanding of the laws or conditions governing the system in question.

Let us compare the genetic example with a probabilistic functional law that shows how a driver's blood alcohol level correlates with car accident rates. It could predict the likelihood of an accident in the next twenty minutes, given a specific driver's blood alcohol level. However, it cannot predict the impossibility of an accident in the absence of alcohol in the blood. In this case, an actual necessary condition for a car accident must be riding in a car, for example. The alcohol level also cannot cause a necessary condition; therefore, it is also not a permissive condition[8] (as I explain later). On the other hand, with two platforms in the ball's trajectory (Dray's example), a player cannot double his chances of catching it by climbing both. However, if someone increases their blood alcohol level, it increases the chances of an accident.

That is why I call these types of conditions (alcohol to accident), following the tradition, "contributory conditions." They increase the probability of R in the context of B, but are neither sufficient, nor necessary, nor permissive conditions for R.

There are two types of probabilistic laws, each with its own implications. The first type are mere statistical generalizations, like the example of alcohol's effect on reflexes. They are temporary placeholders for strict laws, reflecting the limits of our knowledge about the background conditions and concurrent forces. The second type, fundamental laws of quantum physics, describe (possibly) essential statistical laws of nature. However, these metaphysical (non-epistemic) probabilistic dispositions or capacities are not readily accepted by deterministic philosophers of science, as they seem to lack a solid foundation in sufficiency strict laws.

From the first kind of probabilistic law, the "step laws," we can never derive a probability value of exactly 0 or 1. That is why sufficient laws are still on the market. And that is why PLs should enter it.

Synthesizing, generally, these two types of laws, the possibilist and the probabilistic, describe the causal influences of two very different types of conditions: the necessary, whose knowledge gives us the power to avoid a type of event, and the contributory, whose knowledge gives us the power to increase the chances of a specific type of event happening. It makes no logical or methodological sense to describe SL or PLs in terms of probability, as it only complicates its statistical treatment and practical testing. Moreover, most of the time, every time a probabilistic law is generated to describe the work of a contributory condition, it cannot derive possibilities and impossibilities. Only a PL can do so.

For those reasons, I treat PL and SL as distinct from, rather than subsets of, probabilistic laws. Where probabilistic modeling is appropriate, it remains indispensable, but where the explanandum involves structural necessity or logical exclusion, a possibilist formulation provides the sharper and simpler tool.

2.6 Final Remarks

After an overview of the contemporary metaphysical debate over laws of nature, the chapter has broken down the properties and logical forms that laws must possess, emphasizing their role in scientific inquiry. It established that natural laws, regardless of the specific metaphysical framework, share a standard logical structure often represented as a universal quantification of a material implication.

Then I have presented, in detail, the concept of PLs. Unlike SLs, which describe conditions that are sufficient to produce some effect, and probabilistic laws, which express likelihoods of some effect given certain conditions, PLs state necessary conditions for some event. This distinction is fundamental and opens new paths for understanding and predicting phenomena, particularly in contexts where traditional deterministic or probabilistic laws are not available.

I have also addressed critiques of laws of possibility, such as their perceived unfalsifiability, by showing how these laws can be predictive and, therefore, falsifiable. The formalization of PLs, along with the introduction of impossibilist laws, adds a novel dimension to our comprehension of scientific laws, offering unique and irreplaceable kinds of predictions. That is why PLs do something probabilistic laws cannot: they deliver modal verdicts (strict impossibility) and reveal the manipulable conditions behind those verdicts. Now, Chapter 3 formalizes possibilist explanations and addresses the objections of Dray and Reiner.

3

Explanations and Possibilism

This chapter begins with my definition of explanation as the presentation of the reason for the occurrence of something. Next, I will define a specific type of explanation, namely, the scientific explanation. In doing so, at least the existence of two basic kinds of scientific explanations, the "why-necessarily" and the "how-possibly" ones, will be clear. For the first kind, I provide a detailed account of the deductive-nomological model (DNM) model and its criticisms, as well as all the types of causal models, including an introduction to the mechanistic model, which will be detailed in Chapter 5. For the "how-possibly," I will present an overview of all its variations and criticisms. It is essential to make evident all the problems that my account of how-possibly explanation (HPE), the possibilist explanation (PE), solves. Finally, I will define a PE, introduce the concept of a permissive condition, present the necessary and sufficient conditions for a PE, provide examples of its application and situations where it is required, and address some possible criticisms.

3.1 What Is an Explanation, After All?

Since Aristotle, we have called "explanation" pointing out the causes of a phenomenon's occurrence. However, a "cause" may mean many things. He gave four means of it: efficient, material, formal, and final. Nowadays, when science discusses cause, it refers to the first, the efficient, or a sufficient cause for something to change or become something else.

One of those concepts of "cause" was quite different from our scientific use of the term: the final cause. It is the purpose, the aim for which something exists or was done. In his understanding, you can say four things when demanding an explanation of one goddess statue; in other words, when answering why a goddess statue exists: it exists because the sculptor did it, because of the material it is made of (marble), because of the form it has (goddess), or to worship or to contemplate beauty (final).

What Aristotle used to call "final cause" is two different things. Nowadays, we refer to them as "reason" and "function." It is "reason" regarding the agent's motives for doing something and "function" regarding what some object does or can do in the world. In the case of the goddess statue, the sculptor's motivation may have been driven by the need for financial gain from the order, and the statue's purpose may have been to serve as an object of worship.

Since Aristotle's simple definition involves considerable ambiguity and imprecision, I will use a slightly different definition of explanation (the product, not the process). Here, an *explanation is an articulated group of propositions that provides reasons for the occurrence of some phenomenon, taken for granted, in response to a "why" or "how" question*.

In sum, explaining is giving reasons for the occurrence of something.

3.1.1 What Are Reasons?

My definition follows the principle of sufficient reason. This principle states that for everything that exists or happens, there is a reason why it is so and not otherwise. We can trace it back to Leibniz, who defended "that nothing takes place without sufficient reason, that is, that nothing happens without it being possible for someone who knows enough things to give a reason sufficient to determine why it is so and not otherwise" (1716, p. 210).

In this general meaning, "reason" is an umbrella concept that goes for (i) the rational motive of an agent to act in some way, (ii) the function something has in a system, (iii) the cause of an effect, or (iv) a logical implication. Therefore, we should be cautious when using the term. In everyday life, reasons are usually held by only one of those senses: rationales or justifications for a particular decision (a motive). Causes, on the other hand, are physical forces that lead to a specific outcome. As functional explanations are mostly confined to the field of evolutionary biology and logical explanations to logic, explanations typically distinguish between reasons (related to human action) and causes (related to objective physical processes).

Of course, this distinction has its critics. Take, for instance, Donald Davidson (1963), who argues that reasons (motives) can be seen as causes. He posits that when an individual acts for a reason, that reason can be seen as the cause of the action. At the same time, he also defends that reasons are deterministically caused, blurring the lines between these two concepts.

However, upholding the distinction between these two terms is still useful. As I believe that we cannot definitively answer the question of the deterministic origins of reasons through science, reason-based explanations are not the focus of our discussion here. They are not scientific. And scientific explanation is our primary objective.

3.1.2 What Are Causes?

So, we are stuck with "cause." The concept of "cause" could also be messy. Let us remember Hume's classic definition of it: "an object followed by another, where all objects similar to the first are followed by objects similar to the second. Or, in other words, where, if the first object had not been, the second would never have existed" (1952, p. 477).

As Lewis (1973a) has pointed out, Hume's "in other words" leads, in fact, to a second and distinct definition of "cause." For Lewis, the cause of a phenomenon is an event that occurred without which the phenomenon would not have occurred. In other words (this time with similar meaning), for Lewis, an event x is the cause of an event y if and only if y causally depends on x and they are parts of the same causal chain. In his words:

> We think of a cause as something that makes a difference, and the difference it makes must be a difference to what would have happened without it. If it had been absent, its effects—some of them, at least, and generally all—would have been absent too. (Lewis, 1973, p. 161)

We can say that, where B and A are distinct real events, B causally depends on A if and only if B would not occur if A did not occur. Therefore, giving a cause of B may simply be establishing that A was the case. As Glennan (2017) reminds us, we can think of dependence as a class of irreflexive (A does not depend on A), nonsymmetric (if B depends on A, A may or may not depend on B), and transitive (if C depends on B and B depends on A, so C depends on A) relations. These formal similarities apply to any use of the concept of dependence, not just causality.

In ordinary English, this second meaning gives rise to three loose uses of the word "cause," each of which labels a kind of condition. The first is when we say that water causes the growth of a plant. In this case, we mean that water is necessary for a plant to grow, along with other essential conditions such as soil quality and sunlight. Another use is when we say that cutting the plant at its base will cause its death. In this case, we mean that cutting the plant is a sufficient condition to kill it, but we also know that other acts can equally do so, such as pulling the plant out of the ground, burning it, drowning it, or placing it in a dark chamber. Finally, there is the use of the word when we say that only the proper combination of water, soil quality, and sunlight will cause plant growth. In this last case, we intend to say that "cause" is the entire set of necessary and sufficient conditions for something.

As we will see ahead, these different concepts of "cause" will generate different explanations. Generally, when we attempt to explain how an event was produced, we aim to reveal its sufficient condition. When we try to explain why an event was prevented or not, we attempt to reveal an occult necessary condition that was either absent or present. However, sometimes, when we demand a complete explanation for an event, we will try to set all the necessary conditions and the proper sufficient conditions for it. Because whenever an event occurs, at least one sufficient condition and all the necessary conditions are present.

A simple example of this is given by Hurley (2018). The necessary conditions for lighting a match are a certain level of heat and oxygen. Combining these two necessary conditions gives sufficient conditions for the lighting. In other words, striking the match in the presence of oxygen is sufficient to ignite it. A simple and (maybe) complete (all the necessary and sufficient conditions) explanation for the lighting of a match would only include those two necessary conditions that, together, are the sufficient conditions to produce the event.

So, are "condition" and "cause" synonymous? It may appear so if we adopt Lewis's counterfactual definition of cause. However, Lewis himself rejects that conclusion. For him, many factors remain mere "background" conditions: a putative factor counts as a cause only if it makes a counterfactual difference or at least stands in a chain of such differences (Lewis, 1973).

Common sense also draws a pragmatic line between cause and condition. We typically refer to the factor that is abnormal within the total set of necessary and sufficient conditions for an effect as the cause. Hence, we say a short-circuit (spark) caused an apartment fire, but we never single out the oxygen in the air as the cause. Hart and Honoré first analyzed this contrast in *Causation in the Law* (1959), and Mackie (1965) later captured it formally with his INUS analysis.

For Mackie, the spark is a cause because it is an *Insufficient but Necessary part of an Unnecessary but Sufficient* (INUS) condition:

Insufficient by itself: without oxygen and fuel, a spark does not guarantee a fire.
Necessary within one relevant set: given oxygen, combustible material, and no sprinkler discharge, remove the spark, and that fire would not start.
Unnecessary as a complex: other complexes (e.g., a lit match plus the same background conditions) could have produced the fire.
Sufficient as a whole: when the spark occurs together with those background factors, ignition follows.

On views such as those of Hart and Honoré (1959) and Mackie (1965), every cause is indeed a condition, though not every condition is a cause. In this book, however, I adopt Lewis's definition: a cause is any antecedent situated in a chain of counterfactual dependence (Lewis, 1973). In plain terms, a cause is anything that makes a difference in the real world, determining whether the effect occurs. If C is a cause of E, then had C not happened, E would not have happened. Remove C (or some later consequence of C), and the effect E disappears.

By "condition," on the other hand, I take a purely logical relation—something that specifies the circumstances under which an event can or cannot occur. The two traditional kinds are: Necessary condition: if ¬N, then the effect E cannot occur; Sufficient condition: if S, then E must occur.

Nevertheless, in this book, I introduce three further kinds of conditions. The first is the contributory condition. Seldom is "contributory condition" used by philosophers (Alchourrón, 1993; Manero, 2023) other than as a synonym for an INUS condition, but that is not the sense used here. What I call a contributory condition C is an INUS condition for E that is not a necessary condition for E.

For example, in the case of the spark, oxygen is an insufficient-but-necessary element of an unnecessary-but-sufficient set for the fire. It is therefore an INUS factor for the fire. However, in my scheme, it does not count as a contributory condition but as a necessary one, present in every sufficient set for E.

With this specification, I aim for the logical concept of "contributory condition" to more closely resemble the causal concept of "contributory cause," which denotes any factor that helps produce the effect, as the term is used in legal or medical practice. In other words, a contributory condition is a logical construct that identifies a member of at least one minimal sufficient set for E. However, later in the book, I will define another feature of a contributory condition: it is not itself necessary for E.

The remaining two types of conditions, defined later in this book, are the permissive and the preventive. A permissive condition is a sufficient (though unnecessary) condition for some necessary condition of E and will be addressed in this chapter. A preventive condition V for E is sufficient—but not necessary—for the negation of a necessary condition N ($V \rightarrow \neg N$); that is, it precludes E by blocking one of its necessary conditions. This topic will be taken up in Chapter 4.

3.1.3 What Is the Form of an Explanation?

Every explanation has two parts: the *explanandum* and the *explanans*. These terms come from the Latin verb *explanare*, meaning "to explain" or "to unfold." In the explanandum, the addendum *-ndum* indicates an action that must be taken. It became "that which must be explained." In its turn, the ending *-are* in *explanare* means present participle: "that which explains."

Therefore, the explanandum is the proposition that expresses the event to be explained, and the explanans is the structured set of propositions that explain it. In the last example, the *explanandum* would be the proposition "a match was lit," and the explanans "because it was struck in the presence of oxygen."

One of the major disputes about the proper form of an explanation is its relation to an argument. Generally, we think of an explanation as an argument backward. In an argument, we start from the premises we already assume are true and deduce the conclusion from them. In an explanation, we proceed in the opposite way: what we already know is the fact to be explained (*explanandum*), and from it, we try to find premises that would infer it for sure (*explanans*). This kind of backward reasoning is called abduction (contrasting with the deduction of an argument). The proponents of causal explanations disagree with this; for them, an explanation could be a chain of causes (explanans) leading to the fact to be explained (explanandum). We will investigate this dispute within a few pages.

3.2 What Is a Scientific Explanation, After All?

If an explanation is "an articulated group of propositions that gives reasons for the occurrence of some phenomenon taken for granted in response to a why or a how question," what is a scientific explanation?

To begin with, in a scientific explanation, the reason is always a cause (not rational motives or functions). Causes are predictable and law-ruled, reflected reasons are not. Therefore, reflected reasons (rational motives) are not subject to falsification. That does not mean, of course, that there are no such kinds of reasons, only that they are out of the science domain.

Second, when someone asks me for an explanation, I can provide one based on my beliefs and experiences, hypothetical assumptions, or facts, laws, and causes derived from scientific knowledge. This last case constitutes a scientific explanation.

3.2.1 Abductive Reasoning

Let us come back to abduction. In this reasoning modality, we think of an explanation as a backward argument, where we have already fixed a conclusion. This conclusion is usually a fact that has already happened, and we want to know what premise is missing so that, when combined with the known facts and laws, it would render this certain. In other words, in abduction, we stipulate the premise; we do not determine the conclusion. On the contrary, we use the conclusion and the rules of deduction to show that if the putative premise were accepted, it would explain the reason for the conclusion.

For example, a glass broke in the kitchen (*explanandum*). The impact of a ball breaks the glass in the kitchen window. Children were playing with a ball in the yard. If we accept that a ball was kicked at the window, the broken glass would be explained. However, it could always be a stone or a shot that caused the breakage. Abduction is the typical form of reasoning used by scientists, detectives, and doctors, and its product is always a hypothesis.

This cannot be overstated: every explanation is always hypothetical. Suppose that we have an external film of the backyard showing the ball hitting the glass at the same time as it broke. Is the explanation of the anterior paragraph certain? No, it is not. A child inside the kitchen could have hit the glass simultaneously with a hammer, or even a lost shot could have targeted it at the same moment. There are always unlimited hypotheses that can explain an event; if not, conspiracy theories would not be so popular.

The difference between the product of simple abductive reasoning and what we call explanation is that, in explanation, the premise accepted to explain the fact must have independent justification and not be just speculation. In this last case, it would constitute a possible explanation.

The scientific explanation results from abductive reasoning with a scientific (not just empirical, ideological, or spiritual) justification. In other words, if the abductive premise of the explanans is a law or generalization (glasses break if struck by something heavy), or if it is a fact that is subsumed by the generalization (a ball hit the glass), each will have to have scientific justification (the law is tested, and instruments registered the fact). In a scientific explanation, the entire explanans has independent scientific justification. Therefore, a scientific explanation is always about causes, not (reflected) reasons.

3.2.2 Two Basic Types of Scientific Explanation

Since the advent of Laplacian determinism, a coordinated set of sentences is generally only considered a scientific explanation if it points out jointly sufficient causes for its occurrence. That is because, for Laplacian determinism, every event in the universe is sufficiently determined by antecedent conditions subsumed in natural laws, and, therefore, anything other than the explanation of why an event was physically necessary is, at least, an incomplete explanation of it.

Determinism, as we have seen, posits that all future events in the universe are physically necessary and are already determined. However, even if determinism is true, it is a mistake to believe that a consequence of this is that explaining is just giving the causes that made an event necessary.

It is a mistake because if explaining scientifically is (among other things) to point out the causes of the occurrence of an event, then the pointed causes may be necessary or sufficient conditions. In this case, what constitutes a satisfactory explanation depends on the event that requires explanation. Sometimes, showing something without which an event would not have happened only means explaining how it was possible, not why it was necessary.

The two most influential models of scientific explanation, the covering law (Hempel & Oppenheim, 1948) and the causal-mechanical (Woodward, 2003; Craver, 2007), equally assume that offering a complete scientific explanation of an event requires sufficiently demonstrating its occurrence, or at least giving good reasons for expecting it to have occurred. From this point of view, explaining would only be explaining why something necessarily occurred (why-necessarily). Probabilistic law and explanation, in this case, would involve only incomplete expressions of our knowledge about an event that is physically necessary.

However, as we saw, in everyday life, when we demand an explanation, we are sometimes not even interested in the sufficient cause of a phenomenon, but only in some specific necessary condition of it, likewise, in science. When this happens, explaining is to show how-possibly: to show how something that seemed epistemically impossible or unexpected in certain circumstances (to those who demand the explanation) was *possible* (consistent with our basic knowledge of the world) or expected. In this case, it is not necessary, and generally not even interesting, to show why this *should* have happened.

William Dray (1954) was the first to identify this type of explanation, which he called "how-possibly explanation" (HPE), as well as its widespread use in the discipline of history, and defended it as an alternative to the DNM.

Before the end of this chapter, we will offer a synthetic picture of the currently most influential models of scientific explanation and their primary limits. That will provide the background from which PE emerges and its role in the concert of science.

3.3 "Why-Necessarily" Explanations

As I have already briefly defined, why-necessarily explanations are those that point out why, given the conditions in a given state of affairs, the explained event necessarily had to have occurred, or, in other words, that the explanans determined the occurrence of the explanandum. It is to point out the conditions that were sufficient to instantiate the event to be explained. The two most important models of why-necessarily explanation are, as already mentioned, the covering-law (DNM) and the CM.

The first offers a deductive structure where, in its premises (its explanans), there is at least one general law (Strict Law) and a particular condition, and, in its conclusion (explanandum), the event is to be explained. The latter presents itself as a causal chain of events (explanans) that ends in the event to be explained (explanandum).

Something they both have in common is the assumption that the explanans must describe sufficient conditions for the occurrence of the explanandum. The difference, in this case, is that the DNM does this in the form of an argument, whereas the CM explanation is presented in the form of a causal chain.

Defenders of the DNM of explanation believe in determinism, which leads them to believe that explanations have the same structure as predictions. That means that we can only explain a phenomenon that we could have predicted by knowing the laws that govern it and the initial conditions that were sufficient to determine it.

However, determinism is an unfalsifiable metaphysical thesis and is probably false, considering the results of contemporary physics. Neuroscientists and psychologists with little epistemological (and physical) refinement still tend to assume it without any further justification as a necessary assumption of modern science. After all, we can only know something about the object if it has invariable characteristics. As we already saw, for the activity of science to exist, it is only necessary that something in the universe presents at least some regular aspect that can lead to predictions or descriptions. That is a far cry from requiring that everything in the universe be regular in every respect.

The universe does not need to be a realm of absolute necessity to be knowable. Science can have metaphysical and not just epistemological limits, and whether the universe is entirely determined is also a question of fact, not a condition of possibility for modern science.

I will also add to this rejection of the thesis that determinism is a necessary assumption of modern science (P1) the complete definition of a nondeterministic body of scientific knowledge. Part of it was introduced in the previous chapter, with the definition of possibilist law (PL) and its type of falsification. The other part is introduced in this chapter with a PE. However, having argued against adopting a metaphysical belief in determinism as a necessary condition for the scientific enterprise, I would like to evaluate whether this metaphysical belief is better or desirable. To help decide this, let us then briefly evaluate typical forms of scientific explanation.

3.3.1 Deductive-Nomological Explanation

I will focus my exposition of the DNM on the most important aspects of the problems described and the subsequent arguments, rather than bore the reader with yet another summary of this classic thesis in the philosophy of science.

The DNM, initially proposed by Karl Popper (1959/2002) and later refined by Carl Hempel and Fritz Oppenheim (1948), implicitly assumed determinism. It is not surprising that determinists have received it as an ideal expression of scientific explanation. The DNM, also called the covering law model (as it requires the enunciation of at least a strict law that connects to the particular condition stated and "covers" the phenomenon to be explained), tries to avoid the metaphysical character of an explanation (establishing the cause of something), describing it in purely logical terms.

The deductive nomological (DN) explanation can be defined as a deductive argument in which the premises are a set of general laws and particular initial conditions (explanans) that logically imply the fact that occurred to be explained (explanandum). For the explanans in the DNM to adequately explain the event in question, three conditions need to be satisfied. First, all premises must be true. Second, the explanandum must be logically derived from the premises (these two conditions together summarize the "deductive" component of the model). Third, among the essential premises (not irrelevant to the conclusion), there must be at least one strict law, which characterizes the "nomological" component (from the Greek nomo: law) of the same.

3.3.1.1 *Deductive-Nomological Model Formalization and Analysis*

The logical form of a DN explanation can be represented as a valid deductive argument:

$$L1,L2,\ldots,Ln,C1,C2,\ldots,Cm \vdash E$$

where

Ln are the universal laws applicable to the phenomenon.

Cn are the specific conditions or facts relevant to the particular instance of the phenomenon.

E is the explanandum that logically follows.

The necessary and sufficient conditions for a deductive-nomological explanation (DNE) are:

1) The explanans must contain relevant laws and conditions that are true.
2) The explanandum must logically and necessarily follow from the explanans.
3) All laws and initial conditions in the explanans must be required to derive the explanandum. No part of the explanans should be superfluous.
4) All components of the explanans must be empirically testable and have empirical content, which means that they should be potentially falsifiable.

3.3.1.2 *The Difference between Laws and Accidental Generalizations*

Although Hempel tries to avoid metaphysical explanations, we can affirm that for the DNM, a scientific explanation must present the sufficient causes of a given phenomenon, that is, to show what necessarily determined its occurrence. In other words, for Hempel, a scientific explanation is one in which the explanandum is the only possible outcome that can be expected given the facts and laws described.

More than that, he cannot escape a thesis about what would characterize a "law of nature." For him, within the class of true generalizations, we can distinguish between those that are only “accidentally true” and those that are “laws.” Hempel (1965) used the following examples to illustrate the difference between the two types of generalizations: “All gases expand when heated under constant pressure” (p. 338) would be an example of natural law, while “All members of the Greenbury School Board for 1964 are bald” (p. 339) would be an example of accidental truth.

The examples illustrate his defense of Goodman’s (1955) thesis that the feature distinguishing laws from banal generalizations is that the latter cannot support counterfactual and subjunctive propositions. For example, the law on the expansion of gases can support statements such as “If the oxygen in this cylinder had been heated under constant pressure, it would have expanded.” In contrast, the statement about board members in Greenbury gives no support at all to the conditional “If Robert Crocker had been a member of the Greenbury School board in 1964, then he would be bald.” This characteristic can be called “counterfactual resilience,” the thesis that laws are counterfactually resilient, in other words, that laws support their counterfactuals.

Here, Hempel (1965) reminds us that there is a fundamental difference in the explanatory power of common laws and generalizations. The gas law, combined with the appropriate particular conditions (such as that the oxygen in the cylinder was heated under constant pressure), can explain why the volume of the gas increased. On the other hand, the statement about the class of 1964, if combined with a statement like “Harry Smith was a member of the Greenbury School staff in 1964,” cannot explain why Smith was bald.

This law is hypothetico-deductive, the general law of explanans, the predictive counterpart of DNM. For Hempel, a scientific law must point out the sufficient causes of a given phenomenon in the form of a conditional proposition where a certain type of conditions is the antecedent and the type of predicted phenomenon is the consequent: sufficiency law (SL).

3.3.1.3 The Implicit Adherence to Determinism

Consequently, for Hempel, a prediction and a scientific explanation must have the same structure: the same argument explaining why the event had to have occurred predicts that it will occur under the same conditions. The difference is the position in time in which the argument is about the fact: if the fact in focus has already happened, the argument explains it; if it has not happened yet, the argument predicts it.

However, stating that explaining and predicting are formally identical—that is, that we can only explain an event that we could have perfectly predicted—implies that only events under strict deterministic regulation have a possible explanation. That would, in principle, exclude from the field of science events that are not rigidly determined (such as free actions) that can very well exist and whose existence should not be treated as a matter of principle, but rather as a matter of fact, empirically.

In other words, when defending the thesis that the DNM, as defined by Hempel, was the only ideal form of scientific explanation (or at least the only complete one), it was also assumed that every natural phenomenon has sufficient causes governed by laws. That seems perfectly reasonable, but it is not a logical necessity but rather a metaphysical assumption that (i) can (and should) be false; (ii) implies that any phenomenon for which we cannot explain sufficient causes has no scientific explanation. Assuming this thesis, therefore, is an undue metaphysical extrapolation.

The election of these theses (DNM, strict sufficient laws, and symmetry of prediction and explanation) as scientific ideals has less to do with their evident logical qualities than with their metaphysical implications. If science were always successful in using them, Laplace's metaphysical thesis that reality is absolutely determined, in all its aspects, would reign safe and would not deserve serious opposition.

3.3.1.4 Criticism of the Deductive-Nomological Model

However, the reality is very different. Not only are these concepts and models not necessary conditions for science, but science also makes very little use of them. Likewise, the metaphysical thesis that all phenomena in the universe are absolutely determined is dispensable for science, and considering the results of contemporary physics, it is probably false.

If it were achievable, DNM would be desirable in most of science. Its great theoretical strength lies in the type of natural law it adopts and the prediction it implies. When we have a law that can accurately predict the necessary conditions to produce a phenomenon, and we can control these conditions, we have complete control over this aspect of nature. But to what extent is DNM viable?

I believe that laws of nature play a central role in scientific explanation. The problem lies in determining what kind of law is at work and whether its relationship to particular conditions is logical (i.e., always instantiating its results as a matter of deductive necessity) or causal (always exerting the same causal pressure). The DNM accurately describes the structure of many real scientific explanations, particularly in the natural sciences, but it lacks examples in the human sciences. Furthermore, it has limitations that require a review of its terms.

Criticisms of the DNM are of two basic types. The first is that it would be a narrower model than necessary, and the second is that it would be broader. In other words, the first type of criticism is that it would exclude many simpler explanations from the field of scientific explanation, and the second is that it does not provide all the necessary conditions to determine a scientific explanation. Let me quickly evaluate these classical objections to determine my position on them and how the possibilist account hopes to address them. We are talking about the objections of "lawless explanations," explanatory asymmetry, the causal irrelevance that some consistent premises may have for the explanation (Salmon, 1984, p. 249), and the absence of sufficiency strict laws in the universe (Cartwright, 1983). Finally, I will add my critique of the Hempelian form of DNM.

3.3.1.4.1 Lawless Explanations?

It is well known that perfectly acceptable scientific explanations such as "the patient had lung cancer because he smoked for forty years" do not make explicit use of laws. They are based on generalizations from correlation studies, as we do not know precisely the law that regulates the changes caused by cigarette smoke in lung cells. These cases would suggest to some, as they did to Dray (1957), that the DNM was too rigid to serve as a universal criterion of scientific explanation. However, Hempel (1965) himself responds very well to this objection. For him, the point is that this type of explanation implicitly assumes the existence of laws, whether we are aware of them or not. If the explanation were fully spelled out, these laws, or the provisional generalizations that are playing their part, would come into play. The fact that we do not always use them has more to do with the implicit shared content of knowledge of valid laws and generalizations than with a failure in explanation. That is, we often speak in this manner to avoid redundancy.

3.3.1.4.2 Explanatory Asymmetry

The second problem I will summarize here is the explanatory asymmetry in the DNM. Some counterexamples reveal that good predictors of the occurrence of a phenomenon cannot always be part of its explanation (e.g., a seismograph predicts an earthquake but does not explain why it occurred). Likewise, there are reasonable explanations for unlikely events that would not serve to predict them. Let us use Sylvain Bromberger's classic example (cited in Salmon, 1989) to illustrate the criticism here. At 3:00 p.m. on July 4, a fifty-foot shadow is cast from the flagpole onto the flat, level ground. Given that light travels in straight lines (SL), that at that time the sun is 45 degrees from the ground, that the shadow is fifty feet long, and that a triangle with two equal angles is isosceles (SL), then the mast is fifty feet tall.

In other words, it would be perfectly intuitive for the length of the mast's shadow at this time to be explained by the sun's elevation angle. However, the exposed DNM, which precisely satisfies Hempel's requirements, actually explains the height of the mast as a function of the length of the shadow cast on the ground, which is counterintuitive. In other words, this length would be perfectly adequate to predict the height of the mast but not to explain the "why" of its height. This "because" could be that a carpenter made it this way to meet the specification of a commission, for example, but it certainly has nothing to do with its shadow. So, in this case, Hempel's model is too broad, as it allows something that is not an explanation to be counted as a scientific explanation.

However, is the asymmetry problem really severe for the general DNM model? In my opinion, no. It is a serious problem only for the already failed determination of Logical Positivism to eliminate metaphysics from the philosophy of science, as well as for Hempel's thesis that explanation must be a symmetrical relationship, that is, that explanation and prediction are two sides of the same coin. The information that serves to predict a fact before we know it does not always serve to explain that same fact after we know it. However, if one is willing to accept that any explanation needs to use the concept of cause, the deductive model need not—and should not—be abandoned. Moreover, this can be illustrated by remembering that if x explains y, given the relevant laws and additional facts, it does not necessarily follow that y explains x, given the same laws and facts.

To express this causal direction, we can add to the Hempelian model a necessary condition that all propositions of the explanans occur at a time T1 prior to the time T2 of the occurrence of the explanandum. In the case of the hypothesis of the existence of phenomena of simultaneous causation, it would be enough to use the manipulation criterion (Woodward, 2003) to determine the causal direction. However, I do not intend to enter the metaphysical debate here about the temporal precedence of the cause over the effect; I will simply assume this reasonable assumption.

3.3.1.4.3 Irrelevant Premises

The problem of the irrelevance of premises that may eventually be used can be illustrated by the example of Salmon (1984), which involves deriving Kepler's laws from the conjunction of Boyle's law and auxiliary premises. While the derivation is logically valid—Kepler's laws do follow—it clearly fails as an explanation since Boyle's law, which concerns gas pressure and volume, has no causal relevance to planetary motion. This kind of spurious derivation underscores the inadequacy of purely deductive models of explanation and the need for criteria that ensure explanatory relevance. In other words, how do we prevent spurious derivations from passing off as genuine explanations?

Another example from Salmon that does not follow from pure and simple conjunction expresses the problem more strongly. Suppose a man takes birth control pills. A logically adequate DN explanation of why he does not get pregnant can be given by stating that since this man takes birth control pills regularly, and since these pills prevent pregnancy whenever taken, he necessarily did not become pregnant. Although all statements in the explanans are true, it seems clear that this is not an explanation for why this man does not get pregnant. Once again, Hempel's model is too broad and allows a spurious derivation to be counted as a scientific explanation.

Salmon (1984) precisely diagnosed that the problem of irrelevance and asymmetry arises from the DNM's logical positivist attempt not to explicitly reference the role of causality (metaphysical concept) in explanation, defining it in purely logical terms. Assuming that we are talking about causality here, these problems dissolve, especially if considered together with Van Frassen's (1980) thesis that what will determine a relevant explanation of a phenomenon are pragmatic conditions, those conditions and laws that are not self-evident in the context in which the "why-question" (or "how-possibly question") arose. The relevance of a premise is defined by its condition as a cause of the phenomenon on the one hand and the interests and knowledge of the person asking on the other.

These traditional problems with the DNM are weak because, to be resolved, they simply require the admission that a scientific explanation is not merely a logical relationship but expresses a real causal relationship. However, this does not happen with the following two problems.

3.3.1.4.4 Are There Sufficiency Strict Laws?

As we have already seen in the laws chapter, Nancy Cartwright (1983) pointed out a potentially fatal problem for the feasibility of DNM. That is because it depends on general laws of nature that are generally unavailable in scientific stock. Although we work with a mental model of strict, sufficient laws, we generally have only ceteris paribus laws (CPL). These idealizations do not generate universal predictions in the real world; they only function under ideal conditions. Thus, we can never be sure that the DN explanation of an event (such as the formation of a rainbow) follows deductively from a law. Some may be quick to think that CPLs are works of "pseudosciences" like economics or psychology. But these types of laws dominate the entire scientific scene, from physics to economics (Cartwright, 1999). Where, then, could we see DNM strictly applied?

As Cummings (2000) states, it is well known that the DNM in its original form is vulnerable to a wide variety of counterexamples, which is why it is practically no longer defended in the philosophy of science. However, its central belief—that scientific explanation is sufficient determination—remains extremely widespread among scientists. Cummings considers this a scandal: Why do so many unite faithfully under a model that, despite inspired defenses by "ingenious believers" (p. 1), has already proven indefensible and has a miserable track record of success?

3.3.1.4.5 Are the Laws of Nature Possibilist?

The answer I have to this question is what I defend as the fifth major problem of Hempelian DNM. In part or entirely, the scarce knowledge of strict laws mentioned above may occur not because of our epistemic limitations but because of the nature of reality. The DNM, defined by Hempel, is more than just a logical model of explanation. It is metaphysical question-begging under the guise of logical necessity. It is defended as a scientific ideal despite its practical bankruptcy precisely by those who have a dogmatic relationship with determinism. The structure of the symmetric DNM based on SLs presupposes it. The simple problem with this is that, as I have already argued here, perhaps the universe is different from that, and despite this, we can still investigate it scientifically.

Salmon (1984) shows that what counts as an acceptable scientific explanation depends on the temporal and causal structure of the world, not just the logical structures of the argument. What the world is determines the form of explanation, not vice versa. Moreover, a reasonable metaphysical possibility to be considered is that there are laws in nature that do not prescribe the behavior of certain phenomena but simply restrict their possible manifestations; that is, they limit what is allowed to happen but do not determine what is necessary to happen. I will not further defend this thesis now because this entire book is written to do so. We will still return to this problem. For now, it suffices to recognize that the DNM does not work perfectly by rejecting the concept of cause.

3.3.2 Probabilistic Explanations

In the realm of scientific explanation, probabilistic models provide a framework for understanding phenomena where strict deterministic laws are inadequate. Their *explanans* simply needs to offer good reasons to expect the *explanandum* to have occurred, assuming gaps in our knowledge of the set of conditions that necessarily and sufficiently caused the event to be explained. This subitem examines two primary models: the Inductive-Statistical (I-S) model and the Statistical-Relevance (S-R) model, exploring their strengths, weaknesses, and philosophical implications.

3.3.2.1 Inductive-Statistical Model

The I-S model is the probabilistic version of Hempel's DNM, designed to explain phenomena governed by probabilistic (statistical) laws and generalizations, rather than SLs. Hempel always knew that "not all scientific explanations are based on laws of strictly universal form" (1666, p. 58).

He gave us the example of little Jim, who got the measles. For Hempel, it is a perfectly acceptable scientific explanation to say that Jim "caught the disease from his brother, who had a bad case of the measles some days earlier" (p. 58). The explanandum here is Jim contracting measles, and the explanans concerns the earlier occurrence of Jim's exposure to measles. What is missing in my description is a law. However, there is a connection between exposure to measles and contracting the disease. That connection cannot be adequately expressed by a sufficient law, as not every case of exposure to measles results in contagion. That is when Hempel's words make this explanation probabilistic. Because everything we can say is that persons exposed to the same environment and time as people with measles have a likelihood (say 60 percent) of infection given exposure, nothing more.

Hempel calls the result of this reasoning "argument," even though it does not deduce the conclusion but instead induces it. Importing symbols from informal logic for inductions, as the double line indicates, "makes probable in the rate above." He schematized this explanation as:

(i) The probability of people exposed to measles catching the disease is high.
(ii) Jim was exposed to the measles.
================================== [makes highly probable]
(iii) Jim caught the measles.

A more strict formalization of such an explanation would start with the form of a probabilistic law $\forall x\,(C(x) \rightarrow P(\text{Outcome}|\text{Condition})=p)$, with the function P giving a result between 0 and 1 (p), which means the proportion in which the expected result (Outcome) happens when the condition C happens. It would go like this:

(i) $\forall x\,(C(x) \rightarrow P(R|C)=p)$
(ii) $C(a)$
(iii) $\vdash P(R|C)(a)=p$

That, however, is not a formalization of the informal "argument" presented by Hempel as perfectly acceptable. Suppose the probability of R (catching measles) given

C (being exposed to measles) is 0.6. It means that 60 percent of persons exposed to (some level determined to) measles catch it. The formula above shows that a logical consequence of (i) and (ii) is that a person exposed has a 60 percent chance of catching measles, not that Jim got it. Again, little Jim catching the measles is not a logical consequence of the probabilistic law plus exposition.

We are considering here the problem of explaining individual behaviors based on probabilistic laws sufficiently (at least approximately). Moreover, the answer to this problem is: not possible. Explaining why an event had a high probability of occurring is not the same as explaining why it occurred. That does not explain why little Jim is in the 60 percent unlucky and not the 40 percent lucky who do not contract the disease.

The conclusion is that the I-S model may help explain the behavior of large samples but is useless for individual behavior. Despite its apparent indeterminism, probabilistic explanations (not laws) aim to explain why something happened by subsuming conditions and this event to a general (probabilistic) law. Therefore, this is an incomplete explanation of sufficiency, which assumes that everything is absolutely determined and that the probabilistic law used is only a temporary substitute for our ignorance. The background presupposition of determinism remains.

That is one of many problems with probabilistic explanation, but it is the one that interests me more in my global argumentation. One of the other problems is the so-called "threshold problem" (Salmon, 1971), which concerns what constitutes a "high" probability. It seems evident that Hempel's "solution" is a kind of workaround that necessarily leads to a subjective swamp. As it was thrown, "high probability" is nothing more than a fuzzy concept, bogged down in vagueness, because it is inherently vague and does not specify a clear cutoff point.

Another problem also emphasized by Salmon (1984) is relevance. The I-S model inherits the necessary requirements of the DNM about the relevance of the terms of the explanans. However, the problem of determining a factor's relevance when not even the explanandum can be inferred with certainty is even worse than in the original. Especially when we consider complex systems where probability predictions are used and where numerous variables may influence the outcome. From the first lessons in statistics or methodology, we heard the motto, "Correlation is not causation." We must learn that not all statistically correlated factors are causally (and therefore explanatorily) relevant. To address this problem, Wesley Salmon has developed the S-R Model.

3.3.2.2 Statistical-Relevance Model

Salmon's (1984) S-R model explains events through statistical relevance relations rather than probabilistic laws. The I-R model explains an event by subsuming a general law that predicts its occurrence in the presence of certain conditions with a "high probability." So, if 60 percent of the people who had contact with the measles get the illness, and if Jim had contact with his infected brother, then Jim has a "high (?) probability" (but not certainty) that he got the illness, assuming no other relevant information is provided (he was with surgical mask, he was vaccinated against measles, etc.). Differently, the S-R model does not rely on general laws; it considers all statistically relevant factors, whether they are derived from a law-like generalization. It helps in understanding cases where many conditions interact nonlinearly to affect probabilities.

To the S-R model, an explanation is a set of statistically relevant factors that increases the likelihood of the occurrence of an event. For example, smoking is statistically relevant to lung cancer because it significantly increases the probability of developing lung cancer. However, we know that not everyone who smokes develops lung cancer. This approach to probability makes the S-R model particularly useful in fields like epidemiology and sociology, where causal relationships are extremely complex.

Let us get formal. What precisely means to be "statistically relevant"? A factor F is statistically relevant to an outcome O in a reference class R if the probability of O given F within R, $P(O|F \cap R)$, differs from that of O given R, $P(O|R)$. That means F alters the likelihood of O when known alongside R ($P(O|F \cap R) \neq P(O|R)$). By "reference class," we understand a class of individuals or events that are homogeneously grouped for statistical analysis.

The translation is that for an event E to be explained in the S-R model, it is necessary to identify a set of factors {F1, F2, … , Fn}such that each factor is statistically relevant to E within R. The explanation must show all the factors used to increase or decrease the probability of E occurring relative to the baseline probability of its occurrence in the reference class. We can offer the formula:

$P(E|F1 \cap F2 \cap \ldots \cap Fn \cap R)$ is maximized relative to $P(E|R)$

That means that the predictions of the incidence of effect E, given the intersection of all the relevant factors F with the reference class, are the most accurate possible in relation to the incidence of E in R alone.

There are also several criticisms of the S-R model. One that may have become obvious in my brief and flawed presentation is the obscurity of the concept of "reference class." Its definition for explaining some event can vary significantly in terms of its probability. Moreover, while identifying statistical correlations, real causal relations and causal mechanisms in real-world situations are being neglected. The model built to address the problem of relevance in probabilistic explanations was unable to deliver a solution. On the contrary, the requirement to consider all possible relevant factors creates unfeasible explanations, where the best (and simplest) explanations should be causal mechanisms.

Probabilistic models, whether I-S or S-R, offer valuable insights into statistical regularities and relevance in scientific explanations. They are particularly useful in dealing with complex phenomena. However, we should always be careful about taking statistical correlations as causal relations. Moreover, their limitations in explaining individual cases and reliance on subjective definitions, such as "high probability," expose their essential provisional status. That led us to causal explanation, a more important explanatory field to consider in relation to possibilism.

3.3.3 Causal Explanations

Given the obvious intuitive relationship between explaining and pointing out causes and the dead ends of the anti-metaphysical model of the Hempelian DNM, the idea that scientific explanation must be a causal explanation has reached remarkable consensus in the contemporary philosophy of science. This conclusion is not exactly new since Aristotle. Most of the time when we demand an explanation about a phenomenon, what we demand is to know what caused it. Questions like "Why did he die?" and "What was the cause of his death?" are practically identical.

In fact, this does not necessarily imply discarding the covering-law model, as deducing the occurrence of a phenomenon from a general law, in most cases, is just giving its cause. However, the deductive-nomological and causal models are not completely equivalent; in some cases, they diverge.

Moreover, the main cases of this divergence occur precisely where the DNM fails, as in false symmetries. Returning to the problem of the mast's shadow, in a causal explanation, we say that, given the laws of light, the height of the mast is the cause of the length of the shadow and not the other way around. Causality is an asymmetric relationship: if x is the cause of y, then y is not the cause of x. If explaining a phenomenon is pointing out its causes, as causality is asymmetric, the explanation must also be asymmetric. And as we saw, it is.

3.3.3.1 "Local" Models of Causal-Mechanical Explanation

It was the same Wesley Salmon who, at the beginning of the 1970s, had defended his statistical relevance model of explanation (Salmon, 1971) who, in the mid-eighties, making some of the most careful criticisms of the DNM, abandoned the attempt to characterize explanations or causal relationships in purely statistical terms and initiated (Salmon, 1984) the explanatory tradition that today dominates the scientific explanation debate: the CM model of explanation.

One of the greatest exponents of the CM model, James Woodward (1989), said that in it a causal process is characterized by the ability to transmit a "brand" continuously through a causal interaction. A mark would be a local modification of the structure of a process (which can be an object), such as a dent in a car's bodywork caused by a crash or the modification of the structure of a molecule that reacts with another. A brand is transmitted in a causal interaction, which necessarily involves a space-time intersection between two causal processes that modify the structure of both (Kaiser, 2012); that is, each process starts to have characteristics that it would not have without this interaction. For Woodward (2019), "A process is capable of transmitting a mark if, once the mark is introduced at one spatio-temporal location, it will persist to other spatio-temporal locations even in the absence of any further interaction."

Nevertheless, what would a noncausal process be? A typical example that CM advocates offered is that of the shadow of a moving object. The shadow would be a "pseudo-process" because it would not have the ability to transmit a brand. If we try to modify the shadow by changing the light source, this modification will not persist over time unless we continue to maintain this change in the light source while the object continues to move. In other words, the "mark" will not have been transferred to the shadow structure, as in the case of a legitimate causal process.

A CM explanation, therefore, offers, in its explanans, a part of the causal chain of some event E (explanandum), inventorying part of the causal processes and interactions that led to it. To explain with a CM, as Salmon (1984) said, is to fit event E into a "causal link."

However, the question remains: What does "fit" mean, and what types of "causal interaction," "causal processes," or "marks" should appear in an adequate explanation of event E? While intuitively simple, the CM explanation does not compare to the clarity of the DNM, with the latter's formal specification of necessary and sufficient conditions.

By its primitive intuition of "mark" and its contiguous spatiotemporal transmission of action by contact, mechanics is the quintessential field of comfort for the CM model of explanation and the source of its most examples. The pool table comes into play as is typical when we want to give examples about simple deterministic processes. It is no coincidence that Salmon (1984) and Woodward (2019), more than three decades later, resort to this classic image of mechanics to illustrate the CM explanation. The idea is that the explanation of the movement of a previously stationary black ball is given by the description of the causal process initiated by a bat set in motion hitting a white ball that is set in motion, and, upon hitting the black ball, causes its movement (just as it changes direction). The impact of the bat also transmits blue chalk to the white ball, which is then transferred to the black ball on impact. The bat, the white ball, and the black ball are causal processes; the chalk remains are a "mark," and the collisions are causal interactions. For the CM model, citing this causal chain explains the movement of the black ball after the collision (explanandum). The shadow that one ball casts on another due to the ambient light located behind the player does not explain the movement of the balls or transmit a mark because it does not interact causally. Therefore, it should be left out of the explanation about the movement of the black ball.

Critics of the approach to the spatiotemporal location of causal interaction (Hitchcock, 1995; Dowe, 1992) remember that the problem of relevance that affected the DNM, the difficulty of distinguishing between laws and accidental generalizations, also affects the CM explanation conceived in this way. Intuitively, we know that the mass times the speed of the balls (the momentum) and not the ball's color or the presence of the blue chalk mark is relevant to explaining the motion. However, Salmon's CM model does not give us a criterion by which we should choose the linear momentum of the balls and not the blue chalk mark as causally relevant to the event to be explained.

In other words, the CM model has to be able to distinguish causal sequences relevant to the explanation of the explanandum from mere coincidences. Just like strict laws, propositions about causal relationships allow generating counterfactual propositions, because like laws, they express some type of necessity.

As Hitchcock (1995) reminds us, this problem is equally relevant to the example originally formulated by Salmon (1984), contraceptive pills against MND. According to Woodward (2019), this would be because the idea of a brand does not help determine which process property is causally relevant in an explanation, but only because there was a causal interaction between two processes.

As if the problem of irrelevance were not enough, we still have to face the simple fact that some physical forces do not act locally but rather at a distance, the paradigmatic case being that of gravity. It is curious to remember that when this theory emerged, it aroused opposition from Cartesian mechanists who could not accept the idea of causation at a distance. In the 1980s, an attempt was made to reestablish in modern science the idea that some type of spatiotemporal contiguity must always explain the cause via the neomechanists who advocated the original CM model.

In an attempt to resolve these two problems, Dowe (1992) proposed a new criterion for determining what counts as a causal process: conserved quantities. According to this model, a causal process is defined as an object's history with a nonzero conserved quantity of any measurable quantity (mass, energy, charge, etc.). A causal interaction would be a crossing of object histories that involves an exchange of some conserved quantity.

This is an approach to causality that physicalists and metaphysical fundamentalists have welcomed. As Craver (2007) reminds us, if causal interactions are exchanges of conserved quantities and if conserved quantities are found only at the fundamental level of reality, then every cause is located at the level of fundamental particles (as metaphysical fundamentalism defends).

I believe that both the concept of "mark" and "conserved quantities" are useless for the problem of irrelevance because, strictly speaking, every physical process is causal and leaves countless marks or "changes quantities." Let us take the shadow, chosen initially as a "false process." It (the absence of exposure to light on the part of a surface that remains exposed) also leaves not one but innumerable marks and changes quantities; the point is that most of them are absolutely imperceptible to our senses. The sun on a book on the shelf is subtly altering its structure and color, while the shadow cast on the book next door preserves it more. Like the grass under a bench in the garden or even the billiard balls at a much lower level. However, since we can measure these subtle changes via increasingly sensitive instruments, we become virtually immersed in an infinity of inconceivably minor marks or quantities of quantities generated or altered by the same causal process and infinitely multiply the problem of irrelevance.

However, criticism continued specifically for Dowe's (1992) alternative. Craver (2007), for example, further argues that several causality claims do not involve physically interesting forms of action at a distance but, at the same time, present causal interactions without the intervention of continuous spatiotemporal processes or energy transfer from the cause for that purpose. These would be the allegations of "negative causality" due to omission and prevention (Craver, 2007, p. 78). Although these concepts are still the subject of significant philosophical dispute, they are explanatory in several cases and have a close connection with possibilist knowledge. We will return to them later.

Another significant issue with the "local" CM model, whether it's based on "marks" or "conserved quantities," is its inability to provide comprehensive explanations for the behavior of complex or "higher-level" systems. This is a pressing concern, as it's not just about sociological or economic explanations but also extends to psychological, neuroscientific, and even physical explanations. These explanations, while not explicitly using spatiotemporally continuous causal processes that involve the transfer of quantities, do imply that such processes occur at a lower, more fundamental level. Therefore, it's crucial to explore alternative explanatory frameworks.

Whether we are talking about the behavior of a demand curve or how the pressure of gases changes with their temperature, it does not seem absolutely relevant to discuss the concrete physical means used to exchange goods (physical money, inserting the card into the machine, cell phone, or old check) or the behavior of each individual gas molecule.

Furthermore, this is not just a problem of relevance but of achievability. If we make strict sense of the two "local" forms of the CM model, we would have to consider explanations of the phenomena above as inadequate because they do not trace the entire chain of continuous causal processes based on the transmission of "marks" or "quantities." In other words, if this model were taken seriously, the explanation for the increase in pressure should be based on the behavior of individual molecules and not on the volume of the gas as a whole. And we cannot take this suggestion seriously. In addition to being unrealizable, this type of explanation would exceed the processing conditions of the human mind and would, therefore, not be informative.

Craver (2007), talking about neuroscience, gives us another example of this limitation:

> Although I know of no explanations in neuroscience that violate conservation laws, very few explanations in neuroscience appeal directly to exchanges of conserved quantities. Such exchanges almost always occur well below the "level" of the causal interactions that neuroscientists care about. As a result, very few causal relationships described in neuroscience textbooks can usefully be regimented by assimilating them to conservation laws. (Craver, 2007, p. 77)

3.3.3.2 Manipulationist Model of Causal-Mechanical Explanation

However, at the beginning of the century, the CM explanation gained new and renewed strength with the emergence of James Woodward's manipulationist approach. In a now classic book in the philosophy of science, *Making Things Happen*, from 2003, he presents the simple but quite enlightening idea that what counts as causally relevant to Y is every X that, when manipulated, changes the value of the variable Y. Put another way: X is causally relevant to Y if, when we want to change Y, we can do so through the ideal manipulation of X. And X is explanatorily relevant to Y if it is causally relevant to Y.

The idea that the causal relevance of a CM explanation can be determined through manipulation was quite liberating for those trying to make the causal model viable, particularly in biology, psychology, and neuroscience. However, as Rosenberg (2011) does, it is important to remember that the manipulationist CM model does not offer a reductive analysis (the set of necessary and sufficient conditions) of causality. What Woodward offers is an epistemic criterion for determining when we are talking about a cause of event E and how the use of the concept of cause can be grounded in modern science. C is a cause of event E if there is some manipulation of C that causes E. Remember that event E is generally nothing more than a specific change in the property of a substance or object.

The term "manipulation" connotes a voluntary change in a variable caused by an intelligent agent. Woodward's approach is evidently inspired by experimentalism and our most primal intuitions of causality, developed through our earliest manipulations of objects in early childhood. However, it works equally with "interventions" or "interactions" not voluntarily provoked by a human agent, which are omnipresent in nature.

X is considered by Woodward (2003) to be an "intervention" in Y if the change in Y was due exclusively because of changes in X, whether or not caused by an agent. This approach becomes particularly well suited to those relationships that science describes between variables that assume quantifiable magnitudes, which are those that are easily subjected to experiments, such as astronomical ones.

A synthesis of Woodward and Hitchcock's (2003) description of what would be an ideal intervention I in a cause C to cause effect E:

a) I is a change in the value of C that changes the value of E;
b) In I, the change in E only happens through the change in the value of C;
c) I fully controls the change in the value of C.
d) I does not change the value of E directly;
e) I is not correlated with another variable V that causes E;
f) I does not change E by changing any intervening variable between C and E;

Those familiar with the experimental method will immediately identify the origin of these conditions for an ideal intervention in the criteria for constructing a good experiment. Evidently, as science is a fallible enterprise and experimental intervention is not always possible, less-than-ideal interventions may be considered without better ones but generate explanations and predictions of poorer quality and reliability.

It is very characteristic to note that Woodward's explanatory and causal approach also explains why we only capture the direction of a causal relationship in science through experiments and never through descriptive correlation studies. In the experiment, we can directly manipulate one of the variables and measure the effect of the other, while in correlation studies, the quantified quantities are in the variables, but what is missing is precisely the manipulation of one of them. Moreover, it is well known that correlation studies can only determine that two variables are corelated, not the direction of that relationship or causality (if X and Y are correlated, X can be the cause of Y, Y of X, W can be the cause of X and Y or X and Y can be in a feedback relationship).

Here are two examples of how the manipulative approach eliminates the asymmetry problems that came from DNM: Intervention with the slope or size of the flagpole will change the shadow, but no intervention in the shade will change the flagpole. Changing the position of a barometer hand will not change the weather, but changing the weather will change the barometer hand.

Equally, we can see that the approach helps with the problem of relevance. Remembering the case of the man who takes contraceptive pills to avoid getting pregnant, we can intervene by suspending the medication to see if the person becomes pregnant. If we were to intervene in addition to trying to perform artificial insemination, we would discover that the person does not have a female reproductive system. Then, our explanation of why he cannot get pregnant would finally become intuitively acceptable.

3.3.3.3 Mechanistical Model of Causal-Mechanical Explanation

We then come to the "mechanical" of CM explanation. The mechanistic explanation is reasonably distinct from the mere causal chains of early models of causal explanation. Some of its defenders even defend the existence of a "new mechanical philosophy," which would be a new structure of the philosophy of science in opposition to Logical Positivism (Glennan, 2017; Craver & Tabery, 2023). That is one of the motives by which the new mechanism and its potential relation to possibilism will be addressed in an exclusive chapter. This mini-item exists just to mark the place of mechanistic explanations in the realm of causal explanations.

3.4 "How-Possibly" Explanations

Then, we return to the discrete tradition of HPEs. Today, HPEs' use is not limited to the field of history. They also have an explanatory role in evolutionary biology (O'Hara, 1988; Brandon, 1990; Resnik, 1991; Forber, 2010; and Persson, 2012), political science (Goertz & Starr, 2003), economics (Grüne-Yanoff, 2009), and even some in quantum computing (Cuffaro, 2015).

However, the term created by Dray has been corrupted mainly in its use. Today, we identify two basic types of interpretation of a HPE and many variations of it.

The first, opposed to a why-necessarily explanation, defines an HPE as the explanation of the possibility of a phenomenon's occurrence, not its necessity. That is the meaning of Dray's model and my own.

The second is opposed to what Bokulich (2014) calls "how-actually explanations," which are explanations of how something really happened. In this second sense, an HPE is defined as a possible explanation for the occurrence of a phenomenon. That is an explanation of how something that happened could have happened, not how something actually happened.

Therefore, we sometimes use the term to designate explanations of possibility and sometimes to designate possible explanations. As we can see, even in the underdeveloped field of HPE, there is a dispute regarding the true nature or meaning of the term. Let us analyze this problem a bit.

When one defines an HPE in contrast to why-necessarily explanations, as Dray (1957) does, one seeks to explain how it is possible for an event that was epistemically impossible at first sight to have actually occurred. The why-necessarily explanation, in turn, explains why it was necessary for some event to have occurred. In both cases, we are dealing with what Bokulich (2014) calls "how-actually explanations," explanations about what actually happened. An HPE defined in this way is intended to be a true explanation of the possibility of an actual event.

What Bokulich (2014) calls HPE differs from Dray's conception. For her (following Brandon, 1990), an HPE gives a potential explanation, an adequate answer to some scientific problem that is not ruled out by known facts. It can be a competing hypothesis or a hypothetical model. In this sense, an HPE does not select the actual mechanism, law, or condition that enabled the event that occurred. For her, an HPE explains simply by presenting a model that is a possible answer about how an event could have occurred or a mechanism that works in a way consistent with our current knowledge. It is a possible explanation of an actual or possible event.

To make this confusing and budding field easier for the reader, I will present a brief historical review of the different versions of HPEs, starting with Dray's foundational model. Afterward, the field of Possible Explanations will be analyzed, differing between Global HPEs, Local HPEs, Minimal economic models (MEMs), and Ontic HPEs. This comprehensive analysis of this general kind of explanation sets the framework to properly understand the limits of the originality of my own HPE model by comparing it with these earlier versions.

3.4.1 Dray's Classic Model

William Dray (1954) identified and named an HPE for the first time, showing its wide use in the discipline of History. He considered it an alternative to the DNM. For him, this type of scientific explanation did not use laws or generalizations.

The fundamental difference between HPE and DNE is the type of question it aims to answer. While DNE works on the "Why-necessarily? Because...," HPE works on the "How-possibly? Because...." In other words, Dray (1954, p. 17) argued that DNEs explain a phenomenon by showing that it had to occur (was necessary) given a specific factual context and laws. Another supposed difference is that Dray believed (1954, 1957, 1968) that since HPEs do not seek to establish why an event must have occurred (the necessity of a phenomenon), they do not need to appeal to laws of nature.

An HPE does not explain a phenomenon by showing that it was necessary in that circumstance but simply by showing that it was not totally unexpected, that is, that it was possible. Dray's model, therefore, fits into the first type of HPE that we saw above: explanations of the possibility of an event. The widespread assumption that only the demonstration of necessity could be considered an explanation was called by Dray (1957, p. 168) "a type of deterministic myopia."

The classic example given by Dray (1954) is a baseball game broadcast on the radio, in which the announcer narrates that a player caught a ball about six meters high in the air from the field. The game's locution, in Dray's words: "It's a long ball through center field, and it's heading to hit the fence high up. The center fielder is below it, he catches it, and the hitter is out!" (1957, p. 157). How could he have caught the ball? A Dray-style HPE might explain that this was possible because there was a platform for the score, and the player climbed it to catch the ball. In other words, the player on the platform made the event epistemically possible, although not necessary.

Someone might ask why the HPE above explains the necessary condition for the existence of the platform and ladder and not many others, such as the catcher having an arm, a hand, the ball having a maximum allowed speed, and so on. Dray (1954, p. 19) responds: "Because these facts successfully refute a presumption, sufficiently reasonable in light of our knowledge of the moment, that the defender could not have caught the ball." In other words, because these conditions are relevant in the pragmatic context of the demand for explanation and its issuance.

In summary, Dray (1957) defines an HPE as an explanation of how an event, which at first glance seemed epistemically impossible to the one demanding the explanation, was, in fact, epistemically possible. In other words, an explanation of how the seemingly impossible event is possible in our current knowledge. Dray's classic position holds that a "how-possibly" explanation is distinct from a "why-necessarily" explanation, but both aim to offer a true explanation of a real event. This statement may seem somewhat trivial, but it is not. And we will need it soon.

Throughout the literature, two significant criticisms (Reiner, 1993) were directed at Dray's model of HPE. The first and most common was stated in different ways by Passmore (1958), Strawson (1959), Dietz (1970), and Reydon (2012). They argued that Dray's HPE could be included in the logical form of a deductive argument that relies on the laws of nature, and therefore, it would only be a particular case of the DNM. That is only a criticism because, as we saw above, Dray believed that HPEs did not need to be based on laws of nature. It is not a challenge to the existence or intelligibility of HPE.

Passmore (1958) argues that HPEs can be presented in the logical format of a deductive argument, which, for him, would make them a particular case of the DNM. Paul Strawson (1959) reminds us that an HPE can show that what happened was not, after all, an exception to a law, which would also reveal its common nature with the DNM. Dietz (1970, p. 617) even considers these to be sketches of Hempelian explanations.

Dray (1968) responded to these criticisms by noting, among other things, that the fact that HPEs can be formalized in the DNM style does not convert them into the Hempelian model, since, in HPEs, there is no symmetry between explanation and prediction. Dray was right. However, his critics were not entirely wrong.

Reydon (2012) shows that it is false to believe that there is explanatory force in arguments unsupported by laws and generalizations. The explanatory force of HPE would lie in the laws and generalizations implicitly shared by the person asking the question and the person explaining. In the case of baseball, the law of gravity and the generalizations that human beings do not fly or that they do not have the strength to jump more than four meters, for example. Thus, all these criticisms share the common defense that there are, in fact, laws and generalizations implicit in HPEs, despite Dray's narrative providing no explanation for these.

However, even if HPEs made explicit the laws on which they are implicitly based, which are the laws and generalizations that those who demand and those who explain equally believe, they would still be explanations of how the event was possible, not of why it was necessary. The extent to which they are genuine explanations depends, as Reydon (2012) states, on the general vision of an explanation's necessary objectives.

The second criticism is more severe because it questions the intelligibility of Dray's HPE. Reiner (1993) shows that pointing out particular conditions as necessary conditions is a mistake since, at most, they are considered in isolation (the platform), one among several conditions (a TV tower that the player climbed, a helicopter that lifted him, the use of a propelled flight device, etc.) that can make the event (catching the ball) possible. A (de facto) necessary condition (if and only if x, y is possible) would be formed disjunctively by listing all the conditions that would make catching a ball six meters from the ground possible. Nevertheless, this set is impossible to establish definitively. There will always be new ways to catch the ball.

Let us review Reiner's (1993) argument:

P1—To explain why an event was epistemically possible, it is necessary to establish the satisfaction of some necessary condition.
P2—Establishing the satisfaction of a necessary condition of a concrete event requires the exhaustive enumeration of all the particular conditions that disjunctively define it.
P3—The particular conditions that disjunctively form a necessary condition are infinite.

Therefore,

C1—Scientifically explaining the possibility of an event is impossible.

Reiner's mistake is not recognizing that a necessary condition does not form from the disjunction of all these particular conditions that make a singular event possible. On the contrary, an actual necessary condition is caused by any of these conditions, which I call here permissive conditions of the phenomenon y. A permissive condition P for event E is a factor whose presence, in a given context, brings about or activates some necessary condition N for E in some particular context (a short-circuit P for an ignition source N).

In Reiner's (1993) argument, the set formed by climbing the platform, a TV tower that the player climbed, a helicopter that lifted him, the use of a propelled flight device, and all other unimaginable ways to catch the ball is the actual necessary condition to catch the ball. However, neither of them nor the set formed by all of them is a necessary condition. They all, given the special context of the baseball game, are only permissive conditions that enable the actual necessary condition of being with the ball within the reach of a player's hand.

Therefore, the premise P5—Establishing the satisfaction of a necessary condition of a concrete event requires the exhaustive enumeration of all the particular conditions that conjunctively define it—is false. What Reiner (1993) calls "particular conditions" are permissive conditions to an actual necessary condition. To properly establish the satisfaction of the actual necessary condition, only two things are necessary. First, the identification of N. Second, the justification that the permissive condition P causes N with near-certainty in the relevant context.

The premise P6—"the particular conditions that disjunctively form a necessary condition are infinite"—given that, comes to be senseless. No particular conditions form disjunctively a necessary condition whatsoever. It is not possible to establish an actual necessary condition extensionally. It could only be done intensionally. If we do that, we have a law or generalization explaining the common characteristic (type) that makes all these particular events (tokens) make the phenomenon possible. In other words, I am claiming that there is always a necessary condition that is realized by all the permissive conditions of a given phenomenon. And we will understand in this chapter how to identify it.

Reiner's (1993) argument is wrong, as we will see next. However, Dray's model could not deal with it. Why? Dray's work is quite original and foundational. I consider this work in part a continuation of it. However, there are several criticisms to be made. One of my central claims in this book is that Dray was wrong in believing (1954, 1957, 1968) that since HPEs do not seek to establish why an event must have occurred, they do not need to appeal to laws. Moreover, I intend to demonstrate here that any explanation, HPE or causal, relies, at least implicitly, on laws. Dray's model included.

Furthermore, I aim to demonstrate, using a PE, that either HPE has a hidden deductive-nomological structure or does not function. By showing exactly what structure and type of law an HPE of the kind Dray was looking for needs to have, which is not sufficient or probabilistic, I end up presenting a complete HPE and an advance in our understanding of the topic, even for someone who continues to believe that implicit dependence on laws (which Dray denies) is sufficient for an adequate explanation.

3.4.2 Possible Explanations

A *possible explanation* is nothing more than a hypothetical explanation for some event, one among other hypotheses that, if true, would explain an explanandum.

HPE has a relevant field of application in evolutionary biology (O'Hara, 1988; Brandon, 1990; Resnik, 1991; Forber, 2010), where it provides a possible explanation for a specific biological characteristic within a particular evolutionary context. It is used here, therefore, in the sense that Bokulich (2014) gives to HPE, and that Reydon (2012) calls a possible explanation.

Forber (2010) considers that what HPEs do in evolutionary biology is to determine the spectrum of candidates for explaining a given phenomenon, showing which particular scenarios should be taken seriously as suitable candidates. It was Brandon (1990) who established this more imprecise and speculative meaning that HPEs gain in evolutionism. For him, "a how-possibly explanation is one in which one or more of the explanatory conditions are speculatively postulated" (Brandon, 1990, p. 183).

Again, it is essential to emphasize that this type of HPE differs from what Dray (1968) intended to propose. Although in his first texts on the matter, Dray (1954, 1957) was somewhat dubious (Reydon, 2012, p. 304) about the distinction between these two different forms of HPE, later Dray (1968) made it clear that we cannot confuse "HPEs" with "possible explanations." While Dray's (1968) strict HPE specifies (not speculates) some initial conditions whose knowledge the person demanding the explanation lacks (conditions that, when established, make it evident that the event was possible), Brandon (1990) and evolutionary biologists speculate about these conditions (Forber, 2010, p. 36). For this reason, Reydon (2012) points out that Brandon's type of HPE is not an explanation but a speculation that tries to construct the set of candidates for the explanation. It is a little more than science fiction. For a differentiation between these types of HPEs, see Persson (2009).

To characterize Dray's model of HPE, I used the formula "a true explanation of a real event," underscoring its common nature with why-necessarily explanations. Now, I will enhance this formula. Dray's model aims to offer *a true explanation of the possibility of a real event.* In its turn, Brandon's "possible explanation" aims to be *a possible explanation of a real event.* Let us use the transformations on this formula to help understand the differences between definitions and models of HPEs.

3.4.2.1 Global and Local How-Possibly Explanations

Patrick Forber (2010) presented another model of "possible explanations" in evolutionary biology that distinguishes between global and local HPEs. His objective with these models is to understand how scientists generate and assess hypotheses about evolutionary processes. Let us examine these two models separately to clarify their distinctions from PE.

The global HPEs focus on idealized populations and how some potential process can produce evolutionary change. It is a theoretical exploration of abstract models. These explanations aim to provide a possible mechanism by which an evolutionary phenomenon could occur.

They are not restricted to a specific, real biological instance or event, but attempt to explain a general class of phenomena across different contexts and species. In other words, it looks for an explanation that is theoretically acceptable across a wide range of scenarios, rather than one that explains localized empirical data from specific biological instances.

Global HPE works as follows. Consider an idealized population P, possessing a set S of idealized properties. If an evolutionary model M that utilizes these properties to simulate a process exists, and if this process has a nonzero probability of producing a particular evolutionary characteristic x, then we would have a process that explains the possibility of x.

This type of explanation specifies a set of high-level constraints—such as genetic variability and environmental conditions—that influence evolutionary products, along with variable parameters that can be adjusted within the model. Through computational simulations, the model explores various possible processes by which the population P might acquire the characteristic x.

The objective of Global HPEs is theoretical plausibility rather than empirical probability, as they investigate a wide range of evolutionary scenarios that are possible, albeit not necessarily probable. This approach helps the refinement of theories through successive iterations of modeling and simulation. Following the pattern of my comparative sentences, we can say that Global HPE is a *possible explanation for the possibility of a possible event.*

Yes, the sentence is not wrong. Let me unfold these layers of possibility. By saying "possible explanation," it highlights that a global HPE is not asserting what happened, but what could conceivably happen. It is speculative. By saying "for the possibility," it emphasizes that the explanation is about understanding how something might be possible, and not why it is. Finally, by saying "of a possible event," it underscores that the event in question is not assumed to have happened nor to be likely.

And about local HPEs? They are fewer, but still speculative. They show how "some known process produces, in a manner consistent with the set of local information for a real population, an observed outcome or evolutionary pattern" (Forber, 2010, p. 34). It is an application of models to real biological systems (differently from global HPEs).

Local HPEs are focused and tied to specific, real biological cases or scenarios. Their objective is to demonstrate how a particular evolutionary trait could have occurred under the specific circumstances of a particular environment or organism. These explanations are often supported by detailed empirical data related to the trait to be explained.

Suppose a specific population P, characterized by a concrete set S of observed properties. Moreover, suppose we have an evolutionary model M, constructed to fit these specific properties, which simulates a process with a nonzero probability of leading to a particular evolutionary characteristic x, within this population. In that case, we have a local HPE.

This type of explanation articulates detailed constraints—such as specific genetic traits and environmental factors—that directly influence the evolutionary trajectory of P. Through simulations or empirical data, this model investigates specific processes by which P might have developed the characteristic x.

Local HPEs emphasize empirical grounding and specificity, exploring how theoretically plausible processes might actually occur under specific, well-defined conditions. It is *a possible explanation for the possibility of a real event.*

Forber deduces possibility from probability, offering, as he puts it, new types of just-so stories. His account seeks to identify possible necessary or sufficient conditions that could lead to some specific result, determining the set of candidate explanations that should be considered in further investigations or confirmatory trials.

3.4.2.2 Minimal Economic Model

Something similar to Forber's global how-possibly is found in another type of HPE used in economics, which Grüne-Yanoff (2009) refers to as the MEM. This type of HPE model does not satisfy the conditions for linking to reality, such as having a true explanandum, similarity to the world in its parts, or adherence to supposed natural laws.

The MEM is a strategic tool used by economists to help illustrate and understand complex economic phenomena with the simplest models possible. The basic objective of MEMs is to identify the minimal set of assumptions necessary to generate a particular phenomenon of interest within a model. However, neither the model nor the phenomenon is expected to be real; they are only possible, speculative.

As Cartwright (2002) reminds us, economists regularly build models that describe not real economies, but small simulations of analogous or simplified economies that bear little resemblance to the real economy we want to explain (who has never heard an explanation of the "banana island" model?). Once this model is built, they study it, manipulate it, and get supposedly deductive results from it, much like a thought experiment, but without the control and experimental groups' structure and testing. The results are not predictions or explanations about the real economy, but economists continue to use these types of models because they would generate understanding about the phenomena they model.

MEMs, although structured as explanations, fulfill a similar function. They simply seek to offer a vision of how it would be possible to achieve a result within a feasible economic structure. These models are not intended to provide detailed descriptions of reality, but rather to highlight the core mechanisms that could potentially lead to specific economic outcomes.

They have a set of features such as (i) simplicity: the models have always had few elements and a simple structure, aiming to use the fewest assumptions possible. They are also (ii) abstract: they simplify and abstract many characteristics of the real world to focus only on essential features of the phenomenon of interest. They work in (iii) isolation: for being capable of isolating the key variables of the phenomenon.

MEMs, due to their high level of abstraction and simplicity, often overlook other important factors in the real world. That leads to the economic models that we know, which are simple, elegant, and do not work outside of videogames. Nevertheless, this model describes another speculative dimension of HPEs, to provide a *possible explanation of a possible event*.

3.4.2.3 Ontic How-Possibly Explanations

Finally, I must cite Persson's (2012) model. He argues for the existence of a third type of HPE, which he calls "how-possibly ontic explanations" (p. 275), distinct from how-actually (Dray) and possible (Brandon) HPEs. Persson's model, which he describes as pluralistic, focuses on what can be considered incomplete causal explanations. These explanations establish the existence of a partial mechanism that has, in fact, played a role in generating the phenomenon in question. In his terms, this is a "partial explanation of how X" occurred (Persson, 2012, p. 275).

Persson (2012) illustrates it with an example about the symbiosis between termites and their intestinal protozoa. He discusses a case where scientists investigate how termites can digest cellulose, a complex carbohydrate found in wood. The explanation initially offered is that protozoa in the termites' intestines help break down cellulose.

That is not yet a complete causal explanation. However, it is more than a mere hypothesis or speculative model; it identifies a verified causal component of the mechanism, even though the overall picture may still be incomplete. The finding of an empirically verified part of the mechanism that produces X differentiates his explanation from possible explanations.

The main differences between Persson's and Dray's model are that the latter aims to provide a complete explanation of how an event was possible. In contrast, Persson's model aims to offer an incomplete explanation of an event, revealing part of a mechanism without which the explanandum would not have occurred. This part of the mechanism, despite having some empirical support, lacks complete support to connect it to a comprehensive mechanism that necessarily produces the explanandum. Furthermore, on Persson's account, the explanandum need not seem epistemically impossible, only still unexplained. In other words, his model offers *an incomplete possible explanation of a real event.*

3.4.3 Possibilist Explanation

All models analyzed in the previous subsection are types of "possible explanations," which, by adopting the term "How-possibly Explanation" coined by Dray, had distorted its original meaning and use. Although they all have their own functions and merits, they do not aim to establish the framework of what I call possibilist knowledge (Castañon, 2021a).

The HPE model initially offered by Dray has a fundamental original intuition about the explanations we give of the world. It realizes that, often, explaining something is just showing that it was possible, without being able or willing to demonstrate that it was necessary. And this type of explanation is particularly accepted for explanations of human behavior because we share the basic intuition that, ultimately, there would be no laws that sufficiently determine human action (and if there were, they would be so complex that we could not know them). If it were possible to offer a model that combined the qualities of the DNM and Dray's HPE model, we would have a more solid alternative for explaining the scientific possibility of an event. That is what I intend to present in the next chapter with the PE.

The model of explanation adapted to it is in several ways different from the HPE models considered above. It is, like Dray's model, both how-possibly and how-actually. It is not a possible explanation. It is intended to be *a true explanation of the possibility of a real event.*

It differs from Persson's model because it provides a complete, rather than incomplete, explanation that eliminates the presupposition of the impossibility of an event. Furthermore, unlike Dray's, it is a fully formalized DNE, not a CM one, although it could be presented that way, but not without loss of explanatory depth. This new explanation introduces new concepts into the debate on HPE, causality, and scientific investigation. It also corrects Dray's HPE model, combining it with some qualities of DNM. He identifies the true type of law of nature implicit in Dray's HPE, as well as a new type of condition previously unidentified within it.

Finally, this new model of explanation also changes the nature of the HPE explanandum. As this form of explanation does not fit into any of the previous categories of HPEs, and since the term "how-possibly explanations" was appropriated to designate possible explanations, I will baptize it with the forgotten term "possibilist" to express what it essentially explains: possibility.

My motivation for developing this new model comes from the belief that laws play a fundamental explanatory role. That can be considered a problem in itself, as many of the most influential contemporary authors in the debate on scientific explanation have abandoned this view (Cartwright, 1999; Woodward, 2003; Craver, 2007). However, this is not to say that CM explanations are incorrect or inferior explanations. As I will argue here, what I believe is that any link in the causal chain in a CM explanation, to be fully explanatory, presupposes a law, even if that law is nothing more than the expression of a causal power (capacity) of a component of this chain. Even if this capacity is itself explained by a more basic mechanism, at some point in this regression to deeper levels of organization, we will reach fundamental levels of reality where only natural laws could offer causal explanations.

Having made that introduction, I will reiterate this step-item here, just to show possible explanations in the realm of HPEs, where they belong. However, as possibilism is the focus of this book and a PE is what matters in this chapter, we will analyze it in an exclusive section.

3.5 How Does a Possibilist Explanation Work?

PEs are demanded whenever an event E appears surprising or even impossible, under current scientific laws or commonsense expectations. They keep Dray's question, "How was E possible?" but answer it identifying two distinct elements[1]:

A PL that states a necessary condition N for E.

A permissive condition that is a specific condition P whose presence is sufficient to instantiate the necessary condition N for the occurrence of E.

Once N and P are made explicit, the occurrence of E is no longer puzzling: N shows what must hold for E to be possible (and apparently did not), while P shows that it did hold in the situation at hand. By identifying a falsifiable PL, a PE escapes Dray's law-skepticism.

Here, I will take the term "epistemically possible" as the property of a judgment that expresses a state of affairs in the world that is consistent with the current state of our knowledge about the world. In other words, a judgment is said to be epistemically possible if it can be true, given everything we know about the world. Thus, in all uses of the symbol $\Diamond$ in this chapter, I will express this sense of "possibility," not the metaphysical or even the physical. It is essential to keep this in mind, because if the possibility operator used in the formalizations is interpreted in a physical sense, the formalizations will appear incorrect.

To make it more precise, a PE is a scientific explanation demanded when an event E is supposedly epistemically impossible in a context S, and yet it is the case that E (there is no oxygen in an isolated chamber, and yet there is combustion). It states the conditions (permissive and necessary) that, given the PL L, eliminated the epistemic impossibility of event E (e.g., there was a failure in the insulation and therefore there was oxygen in the chamber, given that oxygen allows combustion, it was not impossible for it to happen), rather than those that made it epistemically necessary.

One might first ask: why wouldn't sufficient laws explain these cases? While they can sometimes do so, even in such cases, they do so only through some PL deducible from them. However, in most instances, such SLs simply are not available. Scientifically, all one can obtain is a PL, and it will do the job. A PL is the minimal formulation that can account for the phenomenon.

The PE operates not by explaining event E, but by constructing a valid deductive argument based on laws of nature that eliminates the presumption of epistemic impossibility of event E. This argument follows a logical structure: "Given the laws and generalizations X, Y, and Z, and the particular conditions (not later in time) x, y, and z, it follows that E was not epistemically impossible."

Let us continue working with Dray's (1954) classic example of a baseball game broadcast to show how PE makes explicit the scientific laws or well-confirmed generalizations implicitly assumed by its original model. In a context where we believe that some necessary condition for a player (a) to catch the ball (E) is not satisfied,[2] like having it within reach (N), a PE would explain that catching the ball was possible by stating that:

1) If someone (x) does not have the ball within reach, it is not possible for that person to catch it ($\neg\Diamond Ex$). Having the ball within reach (Nm) is a necessary condition (N) for catching it (E).
2) In context S, the player (a) being on the scoring platform in the trajectory of the ball is a permissive condition (P) that allows the player to have the ball within reach of his hand (Nm).
3) The player used the means of being on the platform (Pa).
4) Therefore, the ball was within reach of his hand (Nma).
5) Therefore, it was not epistemically impossible for the player to catch the ball $\neg(\neg\Diamond Ea)$.

If we were talking here about physical possibility, even if the possibility N (x having the ball within reach) was satisfied, another N1 (x not having hand paralysis) could not be, and this would not allow us to affirm that E was possible (or not-impossible). Nevertheless, since the possibility is epistemic, and it is not discriminated that we know that another necessary condition is not satisfied, then, from the point of view of what we know, the possibility of the event (not its necessity) is explained. The formal structure of an EP, therefore, is:[3]

1) $\forall x (\neg Nx \rightarrow \neg\Diamond Ex)$ (scientific possibilist law)
2) In S, $Pa \rightarrow Na$ (in S context, P is permissive)
3) In S, Pa (The permissive condition P is satisfied)

4) ∴ In S, Na (The necessary condition N is satisfied)
5) ∴ In S, ¬(¬◊Ea) (The event was not epistemically impossible)

This reasoning highlights a key aspect of PEs: they are not about proving absolute (physical) possibility but rather about establishing the plausibility of an event given the available knowledge. If we do not have information indicating that any necessary condition to some event is unmet, we cannot rule out the possibility of the event.

3.5.1 Permissive Condition

I used the new concept of permissive condition here without a prior proper definition. Now, I must define it. I will do it fulfilling another promise, the refutation of Reiner's argument. As we saw earlier (Reiner, 1993), Dray's HPE mistakenly takes one of several conditions, P, that can make an event E possible, as a necessary condition for E. Being on the platform is not a necessary condition for catching the ball. Having it at your fingertips is. Pointing out these particular conditions as necessary conditions is a mistake because, at best, each of them is considered separately (the platform), one of several conditions (climbing a light tower, a helicopter lifting the player, flying a propulsion device, etc.) that makes event E (catching the ball) possible.

Let us review Reiner's (1993, p. 64) argument:

P4—To explain why an event was epistemically possible, it is necessary to establish the satisfaction of some necessary condition.
P5—Establishing the satisfaction of a necessary condition of a concrete event requires the exhaustive enumeration of all the particular conditions that disjunctively define it.
P6—The particular conditions that disjunctively form a necessary condition are infinite.

Therefore,

C2—Scientifically explaining the possibility of an event is impossible.

However, Reiner (1993) is also wrong. His mistake is not recognizing that a necessary condition does not form from the disjunction of all these particular conditions that make a singular event possible. On the contrary, an actual necessary condition is caused by any of these conditions, which I call here permissive conditions of the phenomenon y. A permissive condition P for event E is a factor whose presence, in a given context, brings about or activates some necessary condition N for E in some particular context (a short-circuit P for an ignition source N).

In Reiner's (1993) argument, the set formed by climbing the platform, a TV tower that the player climbed, a helicopter that lifted him, the use of a propelled flight device, and all other unimaginable ways to catch the ball is the actual necessary condition to catch the ball. However, neither of them nor the set disjunctively formed by all of them is an actual necessary condition. They all, given the special context of the baseball game, are only permissive conditions that enable the actual necessary condition of being with the ball within the reach of a hand.

Therefore, the premise P5—Establishing the satisfaction of a necessary condition of a concrete event requires the exhaustive enumeration of all the particular conditions that disjunctively define it—is false. The mistake is that what Reiner (1993) calls "particular conditions" are permissive conditions: sufficient conditions to an actual necessary condition for E. To properly establish the satisfaction of the actual necessary condition, only two things are necessary. First, the identification of N. Second, justify that the permissive condition P causes N with near-certainty in the relevant context.

The premise P6—"the particular conditions that disjunctively form a necessary condition are infinite"—given that, comes to be senseless. No particular conditions form an actual necessary condition disjunctively whatsoever. It is rarely possible (in complex social or physical systems) in practice to establish a necessary condition extensionally. It could usually only be done intentionally. Despite Reiner (1993, p. 64) conceding that what he calls "a general necessary condition" could be formed "intensionally, by seeking the common characteristics of the disjuncts and grouping them under class predicates," he quickly dismisses this possibility as too difficult to be attempted.

But it is not. And when we do that, we have a law or generalization explaining the common characteristic (type) that makes all these particular events (tokens) make the phenomenon possible. In other words, there is always a necessary condition that is realized by any permissive condition of a given phenomenon. Reiner's (1993) argument, therefore, is refuted.

Defining directly: a permissive condition P for an event E is a sufficient but not necessary condition for a necessary condition N for E ($P \rightarrow N$). In other words, a permissive condition for an event does not imply the event itself. What it implies is a necessary condition for its occurrence (Castañon, 2021a).

A permissive condition for E:

a) enables E if and only if it is in a particular context S (climbing the platform can only enable E if it is in the ball's trajectory);
b) is an event among others (P1, P2, P3, …) that makes another event E possible, but not necessary ($(P \rightarrow \Diamond E) \wedge \Diamond(P \wedge \neg E) \wedge \Diamond(\neg P \wedge E)$).

Consider the scenario where the player climbs the platform, making it possible to catch the ball ($P \rightarrow \Diamond E$). However, it is also possible that despite climbing the platform, the player fails to catch the ball ($\Diamond(P \wedge \neg E)$). Similarly, it is possible that the player does not climb the platform at all, but is lifted by a helicopter and still manages to catch the ball ($\Diamond(\neg P \wedge E)$).

The introduction of the concept of permissive condition facilitates the strict use of the term "necessary condition" for E ($E \rightarrow N$), avoiding its usual misuse to refer to any *sine qua non* (Hart & Honoré, 1985), INUS (Mackie, 1980), and even "probabilistic necessary" conditions (Goertz & Starr, 2003).

The condition N (having the ball within reach) caused by P is the actual necessary (but not sufficient) condition for E and is not context-dependent (unlike the permissive condition P). If the player caught the ball, then, in every possible world governed by the same laws as ours, it was within reach of his hand ($E \rightarrow N$). If it had not been within reach of his hand, he could not have caught the ball ($\neg N \rightarrow \neg E$). However, he could have had the ball within reach and still not been able to catch it ($\Diamond (N \wedge \neg E)$).

A caveat is required to clarify the significance of the INUS condition concept and its distinction from both permissive and necessary conditions. In *The Cement of the Universe* (1980), John Mackie seeks to identify causes in the real world with INUS conditions. As we saw, the term INUS is the acronym of *Insufficient but Necessary part of an Unnecessary but Sufficient* condition. That means an INUS condition is insufficient to produce the target effect, but it is a necessary part of a larger set of conditions which is sufficient for the effect to occur. Remember that this whole sufficient condition itself is not necessary, because other sufficient conditions may also produce the same effect. In other words, Mackie shows us that a cause is often just one (indispensable) part of one possible set of conditions that can bring about an effect.

Revisit Mackie's classic house-blaze example. Suppose a blaze is produced by an overloaded electrical circuit (condition), dry wood in the attic (condition), and the absence of a fire-sprinkler system (condition). This set of conditions is sufficient for the fire, yet no single member is. On its own, an overloaded circuit is not sufficient to start a blaze (flammable material is also required); however, without it, this particular combination would not be sufficient. Thus, Mackie classifies the overloaded circuit as an INUS condition.

However, that diagnosis needs refinement. An overloaded circuit may, or may not, cause a short-circuit, which in turn will always generate an ignition source (e.g., a spark). An ignition source is a necessary precondition for a house fire because, in every actual blaze, some initial heat point exists. Nevertheless, without flammable material or when active sprinklers are present, that ignition source is still not sufficient. In this setup, the spark is an INUS condition. At the same time, the short-circuit is permissive, though it also qualifies as INUS within Mackie's broader schema, because it is indispensable inside one sufficient complex.

INUS is therefore a broad construct designed to identify causes. Because it collapses several distinguishable roles (producer, enabler, background), it is less granular than the narrower concepts of permissive and necessary conditions, which focus on differentiating roles in causal chains. In the previous example, not only the spark but also the short-circuit—and even the original candidate, the overloaded circuit—can be counted as INUS factors. They are conflated in Mackie's narrative, yet they differ: an overloaded circuit may or may not lead to a short-circuit; a short-circuit always yields an ignition source; and a match or lighter can yield the exact ignition source without any electrical fault.

Hence, a single factor (such as a short-circuit) can qualify as both permissive (because $P \rightarrow N$, it guarantees an ignition source) and INUS (because it is indispensable within one sufficient complex). The two notions mark different logical relations rather than mutually exclusive categories.

We can also imagine a situation in which a permissive condition for a blaze is satisfied but is not an INUS factor. Suppose the same electrical fault produces an ignition source, yet active sprinklers extinguish the incipient fire immediately. The necessary condition (ignition) is present, but the complete sufficient complex is interrupted, so no blaze occurs.

While a permissive condition is defined as a factor that guarantees (P → N) at least one necessary condition for E, an INUS factor is characterized solely by its role inside some sufficient complex; it may or may not generate a necessary condition on its own.

With these clarifications, the reader sees that "permissive" specifies a P → N relation, whereas INUS specifies a part–whole sufficiency relation. A single event can satisfy one, both, or neither, so the terms are not synonyms, though they can overlap. Unlike Mackie's INUS parts, a permissive condition is sufficient but not necessary for N, and never entails E by itself. A permissive condition is a simpler and clearer diagnostic tool, but it does not replace the broader concept of INUS.

3.5.2 Analysis of the Possibilist Explanation

Now, with the concept of permissive condition (P), we can explicitly define PE by specifying the necessary and jointly sufficient conditions for its occurrence:

a) Its *explanandum* must be a proposition that is a logical consequence of the *explanans*.
b) Its *explanans* must have at least one proposition L that describes a law or generalization that establishes a necessary condition N for event E;
c) the laws and generalizations present in its explanans must be used in deducing its *explanandum*;
d) the propositions of its *explanans* either generate testable predictions or are descriptions of observations.
e) Its *explanans* has only corroborated and falsifiable propositions.
f) Its *explanans* has at least one proposition that describes a permissive condition P for E, which happened at a time t′ prior to time t of the explanandum.
g) Its *explanans* has at least one proposition that describes an event e′ that satisfies a necessary condition N for E, and e′ occurs at a specific time t″ which is prior to t.
h) Its *explanandum* is a proposition that describes the denial of the epistemic impossibility of event E, which happened at time t.

3.5.3 When Is a Possibilist Explanation Required?

Three types of events may require a PE. The first is the supposedly epistemically impossible. The second ones are the uncommon ones. The third is those whose sufficient causes can never be determined scientifically. Let us briefly revisit the first type.

Supposedly, epistemically impossible events fall into two basic categories. The first is the absence of an expected effect in the face of its supposed sufficient condition (its nonoccurrence was supposedly epistemically impossible). To illustrate, consider the example of a stone dropped from the fifth floor that does not touch the ground and remains suspended above it. The second is the presence of an effect in the face of the apparent absence of one of its necessary conditions (its occurrence was supposedly epistemically impossible), such as catching the ball in Dray's baseball game.

The explanation of the second category has already been described. However, explaining a supposedly impossible event in the first category implies changes to the supposedly strict law. If there is some causal interaction that could prevent the effect from occurring, we would need to transform it into a CPL, a probabilistic or even possibilistic law. Even so, an explanation of the possibility of the event could be offered, pointing out which causal influence did not remain unchanged or absent, or was probabilistically insignificant but happened, or whose necessary condition, unknown (until now), was not satisfied.

Since all laws of nature, outside of physics (or even physics: Cartwright, 1983), are ceteris paribus, they all provide predictions that are fundamentally possibilist. CPLs only make conditional predictions; in a strict sense, they only predict that, given x, event y is possible, being necessary only if specific (background) conditions C (uncontrolled) remain constant or absent. When an event expected under a CPL does not occur, we need to show which (background) condition did not remain unchanged or absent to explain how this was possible.

For example, how could it have been the case that a driver who drank 200 ml of vodka did not have his reflexes reduced? An EP could say that, given some generalization that predicts the effect caused by a specific psychotropic drug, the fact that the driver took this drug to increase his reflexes before driving made this possible.

Unusual events also require PEs in addition to causal explanations or DN. Let us distinguish these two demands in the same unusual event. Let us imagine an explosion in a residential apartment in Rio de Janeiro, occurring when João decides to strike a match to light his cigar. Since it is normal for people to strike matches and light cigars in apartments without causing explosions, no one is interested in what sufficiently caused the explosion (we already know) but in what made it possible. However, learning about a gas leak in his insulated and refrigerated apartment on a summer day in Rio (plus an PL about the minimum concentration of gas necessary for the possibility of an explosion) would offer us an interesting PE from the event, like what would appear in the sensationalist newspapers. Since a gas leak is unusual and should not occur, someone would probably say that the explosion was caused by the presence of a quantity of gas, rather than by João lighting his cigarette. However, strictly speaking, the gas leak was a permissive condition that caused the necessary condition (for an explosion) of a given gas concentration. If we want a complete causal or DN explanation of the event, we must add the description of the occurrence of the lighting of a match to this. However, in this case, that is just uninteresting.

Imagine applying this format to the story, which is the target of Dray's concerns. We would explain humans going to the Moon in the 1960s by showing that the discovery of specific physical laws and certain improvements in aerospace engineering made the event possible. That occurred by resolving deficiencies in particular knowledge and technologies that prevented manned space flights a few years earlier. Obviously, we cannot determine the long chain of decisions and contingencies that sufficiently determined the event. However, a PE does not need to determine all the conditions, laws, and contingent facts necessary to do so. It needs only (1) to identify the falsifiable and corroborated law that was apparently violated and (2) to find the permissive condition that caused the necessary condition that was thought to be absent. In other words, a PE must show what explains the epistemic possibility of the phenomenon at stake.

As we can see, even if there is no presumption of epistemic impossibility, an PE can be used to explain merely unusual events. In the case of spaceflight, this can be done by demonstrating all the relevant permissive and necessary conditions that we have access to since we cannot show the entire chain of causal factors that sufficiently caused the event. Obviously, the more satisfied permissive and necessary conditions are known, the more comprehensive the explanation will be. PEs are not always complete explanations (they are only when the goal is to explain why an event that actually happened seemed, but was not, impossible). However, they are a type of explanation that we are sometimes interested in and can obtain completely because, although a DN explanation is complete in theory, it is impossible to obtain, at least in the human sciences. And the same goes for causal explanations.

Finally, as important as using PEs to explain unusual events is using them to explain human events that are simply not sufficiently explainable. For example, you cannot explain sufficiently why some child has manifested behavior that requires understanding the quantity conservation principle, but you can explain why this behavior was already possible for that child. Let us come back to it in the psychology chapter.

3.5.4 Possibilist Explanation Under Some Criticism

Even so, PEs can still be questioned on many points. The most obvious may be the idea that merely explaining the possibility of something is explanatory enough. After all, why would someone need to show that something one already knows happened was possible?

I can give at least four reasons why.

First, we want to show this for the same reason that we want to show that something we already know happened was necessary. We want to know why this was necessary. In PE, we do not want to know that this was possible but rather how it was possible.

Secondly, because, as we have already seen exhaustively, sometimes, considering the current state of our knowledge, the event that occurred should have been impossible.

Thirdly, because, sometimes, considering the current state of our knowledge, the event had to happen but did not happen (i.e., we want to explain how a supposed strict law may have failed).

Finally, because the universe can exhibit phenomena that are not governed by necessity but are nonetheless constrained by necessary conditions and regulated by probabilistic laws (such as personal action). That will be the topic of Chapter 7.

The criticism I attempted to address above is closely related to the issue of the relevance of an PE in the context of certain types of events. One might ask, "Can anyone really have a satisfactory explanation of why someone died, showing that it was possible for that person to die?" Well, when you think of a common phenomenon, of course not. However, if a young athlete suffers a heart attack while watching television at home, his or her loved ones will demand an PE. Conversely, if a man is shot six times in the chest and does not die, we may well want to know how this was possible. No one is saying that EPs are universally in demand. This demand depends on how much we think of a phenomenon as expected or even epistemically necessary.

Another common objection to HPEs—which one might also raise against PEs—is that they are incomplete explanations. First, I should point out that PE provides a complete explanation of the possibility of E in a given context when that possibility requires an explanation.

Second, this depends on one's concept of cause. If we endorse a view of cause as the sufficient condition for the occurrence of the effect, PE would be an incomplete explanation of the occurrence of E. However, even in this case, PE would have relevant use and meaning. Nevertheless, if we endorse Lewis's counterfactual causality analysis, EP can be considered a complete explanation, as it establishes a (necessary) condition without which the event would not have existed.

Third, it depends on what constitutes an explanation and the type of question asked by the person seeking an explanation. An PE (Possibilist Explanation) would be incomplete if and only if (1) the type of explanation required in the situation was why an event was necessary, and (2) sufficient strict laws governed the target effect.

Indeed, if the cause of the phenomenon were an act of will by a human agent (in a dispositionalist sense if consciousness has its causal powers), PE would be the most complete explanation that science could offer with a claim to truth, although not a complete explanation of all the causes of the act. In this case, we could only identify necessary conditions for the behavior in question. However, once we attempt to address sufficient causality, we would need to step beyond the bounds of scientific inquiry due to its inherent metaphysical limitations. In other words, science cannot explain the causes of (or predict) an act of free will.

A third possible criticism I can advance here is that, like the problem of permissive conditions for HPEs, a phenomenon usually has several necessary conditions. However, I must emphasize that this is only a problem of relevance for PE. In PE, the necessary conditions that will be considered relevant depend on the pragmatic context of the law that was apparently violated or, at the very least, faced with a highly improbable outcome. When used in common phenomena, as proposed here in psychology, the relevant necessary condition will depend on the knowledge shared between those who demand and those who explain. Furthermore, as indicated here, a PE can also be an incomplete explanation, which is always open to adding more details. If we do not have a sufficient cause of an event, the more necessary and contributing conditions we know about it, the more comprehensive its explanation will be.

3.6 Final Remarks

In this chapter, I defined "explanation" as an articulated group of propositions that provides reasons for the occurrence of a phenomenon in response to a "why" or a "how" question. A scientific explanation, then, is one where these reasons are causes, and the explanans consists only of scientifically corroborated propositions.

It was explored that there are two main types of scientific explanations: the "why-necessarily," which aims to explain phenomena through sufficient causes, and the "how-possibly," which explains how phenomena could have occurred under certain conditions. The general models of deductive-nomological and CM explanations were briefly discussed to highlight their strengths and limitations in our discussion. That has only underscored the need for a pluralistic approach to scientific explanations that accommodates both explanations of sufficiency and possibility.

About "HPEs," the chapter described their function in explaining phenomena that initially appear inexplicable. They were distinguished from possible explanations, which are merely hypothetical.

PEs were introduced as a deductive-nomological form of HPEs, a scientific explanation demanded when an event is supposedly epistemically impossible in a particular context, yet it occurs. It states the conditions that, given specific PL, eliminate the presumption of epistemic impossibility of the (supposedly impossible) event.

Permissive conditions were introduced as a sufficient but not necessary condition for a necessary condition for some event. We saw that a permissive condition for an event does not imply the event itself; rather, it implies that it is a necessary condition for the event. We set its difference from necessary conditions.

Reiner's critique of explanations of possibilities was refuted, and the demonstration of the form of PEs expanded the scope of scientific inquiry to accommodate explanations of possibilities.

Chapters 5 and 6 will provide empirical and scientific examples to illustrate these types of explanations, showcasing their crucial yet underexplored role in science.

4

Mechanisms and Possibilism

This chapter investigates a rising model of scientific explanation: the "mechanistic." We will begin by describing the basic aspects of what is sometimes considered a new philosophy of science, known as "neomechanicism." After this, I will attempt to clarify how the term "mechanism" is used nowadays and what concept it appears to aim for, and provide my definition of "mechanism" by analyzing all the terms used in it. Based on this definition, mechanistic explanation (ME) itself will be defined as a description of a function especially suited for "how" questions. The rest of this chapter shows how possibilist laws supply the modal backbone those complex mechanisms still need, especially to address negative causality, omissions, and preventions.

4.1 A New Philosophy?

We then come to the "mechanical" of causal-mechanical explanation (CME). The ME is reasonably distinct from the mere causal chains of early models of causal explanation. Some of its proponents even posit the existence of a "new mechanical philosophy," which would be a new structure of the philosophy of science in opposition to Logical Positivism (Glennan, 2017; Craver & Tabery, 2023).

This "new philosophy" cannot deny its affiliation with a long philosophical tradition that goes back to the Cartesian "machine" metaphor. Descartes offered in his 1641 "Meditations on First Philosophy" a revolutionary theory about the nature of matter, space, and universe: mechanicism. In a nutshell, mechanicism starts from the idea that the nature of matter is occupying a place in space, in other words, having an extension. There are no qualities in this universe, only geometrical quantities, and the change of the extension between corpuscles over time (movement). Matter can only have changed the direction of its movement through the transmission of movement from other corpuscles. The fundamental constant in physics—that which never changes—is motion. The universe is a grand, perfectly integrated mechanism like a clock, created by God and initially wound by Him.

This conception of the universe may seem too naïve today, but at the time, it was a fantastic achievement, a liberation of the human mind from Aristotelian metaphysics. Alongside Cartesian geometry, mechanicism provided modern science with a new metaphysics and the tools to mathematize space and matter. Its influence spread through time, as seen in Newton's mechanics, with the eighteenth-century La Mettrie's metaphor of the "man-machine," Laplacian determinism, Pavlov's associationist psychology, and finally, with the "new philosophy" of mechanism.

However, in some aspects, "neomechanists" defend ideas that are quite different from the original tradition. I think the most notable distinction is the contrast between the simple and mechanical view of the nature of original mechanicism and the belief in a complex world, reasonably disordered and endowed with regularities of limited scope, defended by contemporary mechanists like Nancy Cartwright (1999).[1]

Unlike traditional mechanisms, which are likened to clockwork, the "mechanisms" of this neomechanicism are, therefore, only partially stable, subject to outcome variations due to an amazingly complex symphony of causal forces in a dispositionalist world. To deal with this, its advocates often employ more contemporary analogies such as probabilistic computational models and networks. With this metaphorical apparatus, they attempt to explain highly complex phenomena from specialized sciences, such as biology and neuroscience.

This "new philosophy" must reckon with phenomena that classical mechanics cannot address, such as quantum indeterminacy and the behavior of complex systems. It does this trying to understand these phenomena in a broadly mechanistic framework, albeit with modifications to accommodate nonlinearities and supposedly stochastic events.

Other striking characteristics of this "new mechanistic philosophy" defended by Glennan (2017) and Craver and Tabery (2023) would be first the thesis that science seeks to understand the causes of natural events, and not Humean regularities or explanations that purge metaphysics such as the deductive-nomological model (DNM) (Cartwright, 1983, 1999; Salmon, 1984 and Woodward, 2003). It seeks to describe how various parts of a system contribute causally to the behavior of the whole (as in molecular biology and genetics).

A second characteristic would be the defense of special sciences as autonomous domains with their own methods and problems (Fodor, 1974; Cummins, 2010; Glennan, 2017). Moreover, researchers in these fields often question whether fundamental laws of nature even exist for complex phenomena in biology, psychology, and economics. For authors such as Fodor (1991a), Cartwright (1999), Cummins (2010), Craver (2007), or Glennan (2017), laws in special sciences are effects of mechanisms to be explained, rather than ultimate truths of nature that explain them.

This lawless view is supposedly sustained by a hierarchical perspective of nature, where phenomena are understood at multiple levels of organization, each with its MEs. Crucially, neomechanistic causality operates not just linearly (as forward chains of causation) but also vertically—with lower-level mechanisms collectively enabling and constraining higher-level phenomena in a pyramid-like structure of dependencies.

This view is closely linked to another contemporary philosophical phenomenon, emergentism. In neomechanicism, interactions at one level (e.g., molecular) can have emergent properties at a higher level (e.g., cellular), but only in a weak sense: these emergent properties can be entirely explained through the organization, interactions, and physical properties of the components at the lower levels. The other option would be a strong form of emergentism, which defends that some emergent properties cannot be fully reduced to or predicted from the properties and interactions of the system's underlying components. These "strong" emergent properties would possess a form of causal power that is not present in the individual parts, as does consciousness, for example, with its properties of intentionality or qualia. Craver even disdains such an idea as "spooky" (Craver, 2007, p. 216), reiterating the commitment of neomechanicism with a very strict version of materialism:

> I do not advocate a spooky form of emergence. It is important to keep several different senses of the term "emergence" distinct. Some philosophers and scientists use the term "emergence" to describe properties of wholes that are not simple sums of the properties of components. Mechanisms are non-aggregates, and so they are emergent in this weak sense. Mechanisms require the organization of components in cooperative and inhibitory interactions that allow mechanisms to do things that the parts themselves cannot do. (Craver 2007, p. 216)

Nevertheless, before establishing itself as a completely new metaphysical and epistemic paradigm, the new mechanicism has more prosaic challenges to overcome. The first is the precise definition of the concept of "mechanism" itself. The second is the proper definition and logical specification of its own type of explanation.

Now, I will need to expand slightly on the analysis of the concept of "mechanism," as this will define the ME and the relationship between possibilism and the latter. If the reader is not particularly interested in the topic, I suggest that they only read the definition of "mechanism" with which I will work from now on at the end of the following item.

4.2 The Definition of "Mechanism"

This item shows and analyzes definitions of "mechanism" offered by neomechanistic tradition. To address the discrepancies and vagueness, I will offer and work with my own definition of "mechanism." Let us see why.

4.2.1 Neomechanistic Mechanism

In a recent entry written for the Stanford Encyclopedia on the concept of "Mechanisms in Science," Carl Craver, one of the exponents of this approach, recognizes that defenders of ME simply gave up on a reductive analysis of the concept and, "instead, they offer qualitative descriptions designed to capture the way scientists use the term and deploy the concept in their experimental and inferential practices" (Craver & Tabery, 2023, p. 5).

Cartwright (1999) argues that what we call "natural laws" are not fundamental features of nature, but rather regularities produced by what she calls "nomological machines": structured systems that give rise to stable phenomena. While the term "machine" may suggest traditional physical mechanisms, we should keep in mind that Cartwright uses it more broadly to refer to any relatively stable arrangement of components—physical, statistical, or institutional—that, under the right conditions, reliably produces regular behavior. She defines a nomological machine as:

> It is a fixed (enough) arrangement of components, or factors, with stable (enough) capacities that in the right sort of stable (enough) environment will, with repeated operation, give rise to the kind of regular behavior that we represent in our scientific laws. (Cartwright, 1999, p. 50)

This conception still reflects a mechanistic stance in a generalized sense, insofar as it grounds scientific laws in the operation of structured systems with productive capacities, rather than viewing laws as universal, exceptionless forces.

According to Bechtel and Wright (2009, p. 119), two recognized authors often cited by those who address the topic, "a mechanism is an organized system composed in such a way that coordinated operations of the components produce effects and have capacities."

Based on Craver himself, we can observe the fluidity of the concept of "mechanism" in contemporary philosophy. In a now classic paper on the approach written together with Machamer and Darden, Craver defines "mechanisms" as "entities and activities organized such that they are productive of regular changes from start or set-up to finish or termination conditions" (Machamer, Craver & Darden, 2000, p. 3). In "Explaining the Brain," a classic on ME, he defines them as "a set of entities and activities organized such that they exhibit the phenomenon to be explained" (Craver, 2007, p. 5). In the aforementioned entry (Craver & Tabery, 2023), as a compromise solution between the three "most commonly cited" definitions of mechanism, it offers four concepts that would be present in any of its definitions: "phenomenon," "parts," "causes," and "organization," and endorses Glennan's (2017) definition of "minimal mechanism," which, according to them, would offer the set of common denominators for all claims of mechanisms in the different scientific disciplines:

> Minimal mechanism: "A mechanism for a phenomenon consists of entities (or parts) whose activities and interactions are organized so as to be responsible for the phenomenon. (2017, p. 17)

Let us evaluate what the four main concepts taken as common denominators in the above definition of commitment mean for Glennan.

Phenomenon: The centrality of the concept of phenomenon to the mechanism occurs because every mechanism x is so only because it is the producer of phenomenon y, the latter determined by the structure of the parts and interactions of the former. Conversely, it is the phenomenon that we want to explain that determines, with the parts and structures jointly necessary and sufficient to produce it, the limits of the mechanism. Following the tradition started, once again, by Jerry Fodor (1968), authors such as Cummins (1983),

Cartwright (1999), and Craver (2007) consider that the phenomenon explained is the capacity of the mechanism. In other words, mechanisms explain capabilities, not unique events. Here, "phenomenon" means the type-level capacity, not an individual token. While a causal explanation explains the particular event of ball seven moving in a specific match, an ME would explain the mechanism of the pool table. The relationship between the mechanism and the phenomenon it produces is regular if the mechanism performs a function. Nevertheless, authors such as Craver and Tabery (2023), Cartwright (1999), and Glennan (2017) want to leave the question open by taking a vague position on the concept of "regular." As Machamer, Darden, and Craver say, "Mechanisms are regular in that they work always or for the most part in the same way under the same conditions" (Machamer, Darden, & Craver 2000, p. 3. My emphasis).

Parts: Every mechanism comprises parts (or entities) composed of a particular structure (they do not simply come together) and are located in a particular region of space and time. Most mechanists believe these parts are mechanisms whose capacities and dispositions are given by the structure of their parts.

Cause: The mechanism is a causal chain, and MEs are causal explanations. This chain of causality comprises the activities and interactions between the parts that compose it and occurs due to the latter's capabilities. Furthermore, as already stated, the capabilities of the parts are explained by the fact that they are mechanisms at a more fundamental level. In other words, "cause" would be a concept derived from the concept of "mechanism."

Organization: Finally, organization is fundamental in distinguishing a mechanism from mere aggregates and guaranteeing the phenomenon it produces. It is the particular configuration of parts, each with its capabilities and dispositions, that produces a specific phenomenon and guarantees the mechanism as a whole its capacity (production of the phenomenon) that the mere aggregate of its parts would not present. A mere aggregate of parts has properties that are nothing more than the simple sum of the properties of its parts. A mechanism exhibits emergent properties due to how it organizes these parts.

I would add to Glennan's (2017) "minimal mechanism" the role of the environment. Neomechanicists usually agree that it is paramount to consider the environmental conditions in which a mechanism produces its activities. Radical environmental changes would prevent the mechanism from producing regular phenomena, such as a massive variation in the ambient temperature for a biological organism. As Bechtel (2009) remembers, environmental factors can significantly impact how humans perceive visual stimuli. It is always necessary to consider not just the internal state and functioning but the particular kind of inputs the mechanism receives from the environment (Bechtel, 2009, p. 557).

4.2.2 Which Mechanism?

Both Glennan (2017) and Craver and Tabery (2023) miss some of these specifications in their minimum definitions of mechanism to guarantee its descriptive universality in the field of sciences, particularly, in this case, social sciences.

The first concerns the possibility of mechanisms being ephemeral and unique processes. These authors state that mechanisms can be both systems that are prevalent over time and processes that occur only once. Processes are a causal chain of entities acting on each other over time. Therefore, it is evident that any mechanism produces its phenomenon through a process, which constitutes a causal chain. However, not every causal chain process produces its phenomenon through a mechanism. When, in their eagerness to encompass the types of historical causal explanations, these authors embrace ephemeral processes—which are not repeated and are not systematic—as "mechanisms," they end up placing under the scope of the concept all the processes that have taken place in the universe since the Big Bang. If a process occurs but once, what scientific predictions can it yield? Why should science relinquish prediction?

The second concerns the regularity with which the mechanism produces the phenomenon. To avoid defining a mechanism as something that performs a mathematical function (algorithm), these authors (Glennan, 2017; Craver & Tabery, 2023) leave open the possibility that a mechanism, in the same environment, with the same input and in the same state, produces more than one output. By also abandoning regularity as a criterion, they increase the emptiness of the concept, which, in addition to all processes, can now encompass all systems. This consequence makes it very difficult to understand how such a type of "mechanism" could support an explanation of the necessity of a phenomenon under certain circumstances. It is no surprise that Glennan (2017), because of this, argues that the whole world is made of mechanisms; after all, it seems that nothing in the natural world is outside the scope of this concept (un)defined this way.

After this vacuity, defending the infinite regress that all the causal powers of the parts are explainable by more basic mechanisms, as Glennan (2017) does, becomes a minor problem. Furthermore, suppose the argument about the causal powers of the parts falls into an infinite regress. In that case, the very definition of "cause" as the result of a mechanism's activity is circular since, as we have seen, all the definitions evaluated here of "mechanism" use the concept of "cause."

The current state of the definition used by the approach is, therefore, untenable.

Here I would like to present what I believe to be the true implicit definition of Glennan and Craver's "minimal mechanism":

> Explicit Minimal Mechanism: A mechanism for a phenomenon consists of entities (or parts) whose stable (or not) activities and interactions are stably (or not) organized to be responsible for causing the phenomenon in a regular (or not) way.

The problem is analogous to the debate raised by John Searle (1992) regarding a more precise concept, namely, the concept of "computation." Searle argues against computationalism that, if we follow the definitions given by Turing, everything is a digital computer. In his words:

> On the standard textbook definition of computation, it is hard to see how to avoid the following results:

1. For any object there is some description of that object such that under that description the object is a digital computer.
2. For any program and for any sufficiently complex object, there is some description of the object under which it is implementing the program. Thus for example the wall behind my back is right now implementing the Wordstar program, because there is some pattern of molecule movements that is isomorphic with the formal structure of Wordstar. But if the wall is implementing Wordstar, then if it is a big enough wall it is implementing any program, including any program implemented in the brain. (1992, p. 208)

As the new mechanicism struggles to define "mechanism" as a system that performs a function or follows a precise algorithm, what does this imply about the value of defining a "mechanism" that does not perform a function and thus has no definite outcomes?

I do not oppose the general intuitions of neomechanists if taken in a weak rather than a strong sense. I believe that explaining scientifically is giving causes, that sciences such as biology or psychology are autonomous, and that most of what we call laws (if any), or at least regularities, particularly in these special sciences, are effects of mechanisms.

Nevertheless, it seems very clear to me that in their effort to offer a definition that encompasses all the most famous self-proclaimed "mechanisms" of biology, neuroscience, and psychology, as well as to argue for the constitution of a "new paradigm" of philosophy of science, Craver and Tabery (2023), Cartwright (1999), and Glennan (2017) end up offering definitions so imprecise and permissive that practically everything that exists in the natural world can be classified as a "mechanism."

I should only dwell here on the debate surrounding this concept for as long as I have already. However, it is challenging to analyze the relationship between possibilism or scientific explanation and such conflated and sometimes vague concepts. Therefore, I must offer another definition to proceed.

4.2.3 An Alternative Definition of Mechanism

Let me define what I will call "mechanism" from this point onward, particularly in my analysis of the ME from a possibilistic point of view:

> A mechanism M is a stable and organized set S of no free-agents' entities that, in a given environment E, performs a specific and finite order of operations O that transforms a change A in E, that informs the system, with S being in the same internal state I, always in the same phenomenon B at the exit.

Let us break down this definition.

Given that M is a mechanism, then the following conditions apply:

(i) M is a stable and organized set S.
(ii) M contains no free agents (intentional agents).

(iii) M operates within a specific environment E.
(iv) M performs a specific and finite order of operations O when E suffers some specific change A.
(v) M transforms a signal from change A that informs the system, through O, into a phenomenon B.
(vi) M always executes the same function producing the same output B within a specified tolerance band, when activated by the same change A in E and initiated from the same internal state I.
(vii) M returns to the same internal state I after operation O.

Let us work on the words, concepts, and sentences.

(i) "*M is a stable and organized set S.*"
Stable: By this word, I mean that the system of parts constituting a mechanism must persevere over time to preserve its components and organization. I do not mean "indestructible," which means that a mechanism can lose one of its non-redundant parts but, if so, cease to be the same mechanism or even a mechanism. Organization: A structured arrangement of entities.
Following the mechanistic tradition, "organization" distinguishes a mechanism from any kind of agreement and guarantees that the stability of this specific configuration of parts functions produces a specific phenomenon (capacity). The condition (i) specifies that M consists of multiple components arranged in a specific and stable configuration. It could have used the word "system" to describe it, but it was better to be more specific.
(ii) "*M contains no free agents.*"
Free agent: An entity capable of initiating new chains of causality, or in a less radical requirement, an entity that acts independently, taking actions based on its own volition or reasoning rather than being determined by external forces.
This caveat (ii) aims to contrast free agents to mechanical or unique substance parts of a physical system. By vetoing agents, it is guaranteed that each component's behavior is determined by the structure and rules governing the entire set, making M work predictably.
(iii) "*M operates within a specific environment E.*"
Environment: It is the conditions, influences, or forces that operate in the surroundings of something, affecting, conditioning, and even determining its existence, integrity, behavior, or growth.
With condition (iii), M's functioning is defined as dependent on its surroundings, especially physical. M's operations are triggered and conditioned by external factors. For example, a cell phone will not function properly in extremely low or high temperatures and will change its behavior depending on the signals it receives.
M hinges on its stability under bounded environmental variation. While M must reliably transform input A into output B (per conditions iv–vii), the environment E is not static. Rather, E admits a range of permissible fluctuations

(e.g., tolerances in temperature, pressure, or signal intensity) within which change A (the input) can occur without disrupting M's function.

(iv) "M performs a specific and finite order of operations O when E suffers some specific change A."

Operation: Set of coordinated behaviors performed to achieve a specific result.

In (iv), it is guaranteed that M executes a well-defined (specific) sequence of steps, clearly delineated and limited in number, which also construes the predictability of M's function.

(v) "M transforms a signal from change A that informs the system, through O, into a phenomenon B."

Phenomenon: Following the Kantian tradition, by "phenomenon" here, we should understand the way something appears to a mind, specifically those kinds of "apparitions" that are taken by observable events, occurrences, or circumstances that can be perceived through the senses or detected through scientific instruments.

Information: What happens when a signal reaches a receiver, enabling it to choose one outcome between a set of possible alternatives. The most basic unit of information—one binary decision—corresponds to accepting or rejecting a single, indivisible signal element (e.g., a "yes/no" choice), the *bit*.

With (v), it is a necessary condition to be a mechanism that M processes an input (change A) into an output (phenomenon B). The relationship between the mechanism and the phenomenon it produces is regular if the mechanism performs a function (here the term "function" has the mathematical definition).

(vi) "M always executes the same function producing the same output B when activated by the same change A in E and initiated from the same internal state I."

Function: Relation between a set of inputs and a set of permissible outputs with the property that each input is related to exactly one output.

Internal State: The specific configurations within a system that determine its current behaviors or outputs.

Condition (vi) sets the invariance of the output B given the invariance of change A in environment E and when the internal conditions I are controlled, underscoring M's reliability. With this condition, a mechanism performs a function.

(vii) "M returns to the same internal state I after operation O."

Finally, with condition (vii), it is ensured that after processing the input signal, M retains the property of reverting to its original internal state, I. This reset feature guarantees the repeatability of M's operations over multiple cycles. In the definition, this characteristic is determined by the conjunction of the word "stable" with the expression "with S being in the same internal state I," defining the internal aspect of this stability.

Summarizing the central claim: a mechanism performs a (mathematical) function. The above definition greatly restricts the class of phenomena that we can classify as "mechanisms." However, it makes the concept precise and capable of sufficiently explaining other phenomena, such as capacities.

My definition is particularly well-suited for deterministic contexts, such as engineering, computing, or physics. However, would it work in highly variable biological or social systems? The answer is: in biological systems, yes; in social systems, never. If "mechanism" refers to the concept that I have delineated, there are no mechanisms in sociology because its "systems" would be composed of free agents.

Regarding biology, my definition could be nuanced to accommodate systems where outputs are not strictly identical but fall within a predictable range of variance. In this case, the "same output" would be something within this range. In other words, a mechanism can be probabilistic about its outputs. This modifier would maintain the possibility of addressing more complex or adaptive systems, which exhibit characteristics of mechanisms but with additional layers of variability and responsiveness to environmental changes.

Nevertheless, it would not make sense to talk about a mechanism that was not deterministic. From this perspective, a non-deterministic mechanism is not a mechanism. When I provide the possibility of a range of outputs to the same input, I am considering our limited knowledge of complex systems. The hope is that if perfectly known, the mechanism in question would be precisely deterministic.

This does not mean that mechanisms work correctly all the time. An inhibitor or preventer may stop the functioning of one of its parts (such as a medicine or a disease), or it may degrade or break down. However, this indicates, strictly speaking, (1) that the environment E changed and changed the internal state I, or (2) that set S no longer exists and, therefore, neither the state I.

Furthermore, we may need more mechanism hypotheses that do not include some part or activity that does not allow for an accurate prediction of the output phenomenon. I will, however, still argue here that the description of a mechanism offers us several possibilist laws and opportunities for possibilist explanations (PEs) of individual behaviors.

4.2.4 Machines

Some might object that I am reducing the concept of "mechanism" to the concept of "machine." Craver (2007, p. 4), for example, believes that we cannot simply state that a mechanism is a machine because (a) this definition would restrict its use in neuroscience, (b) machines are divisible and have easily identified parts, and (c) machines (such as automobiles, for example) are made up of many mechanisms and also have non-mechanical parts (such as the bodywork). We can also add to these objections that (d) machines are devices made by human beings. At the same time, mechanisms can be products of biological evolution, and (e) machines are capable of active and passive states (he cites the stopped clock as an example). In contrast, mechanisms have a productive aspect and are always active (Craver & Tabery, 2023).

I do not intend to equate the concept of "mechanism" with that of "machine." Nevertheless, not all of Craver's objections strike me as appropriate. Let us briefly evaluate them one by one.

(a) The function of a good definition should be to prescribe the precise use of the term and not to cover all the particular uses it has in the language. In this way, Craver's objection (a) to equalizing the concept of "mechanism" to the concept of "machine" seems simply an attempt to preserve the possibility of fitting any neuroscience model into this category.
(b) The minimum definition that he defends in a later work (Craver & Tabery, 2023) takes it as a necessary condition for a mechanism to be made up of identifiable parts.
(c) If this were the only distinguishing aspect between "mechanism" and "machine," it would be enough to define a mechanism as a "single machine without non-mechanical parts."
(d) I believe that here, we have a defining aspect of the machine concept, which I would rectify as a "mechanism built by an intelligent being to perform a function."
(e) It makes no sense to say that machines are capable of active and passive states while mechanisms are incapable of passive states. Perhaps, when stating this, Craver had biological mechanisms in mind. However, even these, like the mechanism of a genetic disease, can remain passive until activated by a specific input. The watch will only become active when it is wound. No one would object that the clock is, or has, a mechanism that may or may not be producing the phenomenon at that moment.

Thus, to end this long parenthesis from the main book's purpose in this chapter, a definition of machine is offered that distinguishes it from the more general category of mechanism on which it depends.

> Machine: A mechanism built by an intelligent being to perform at least one function, *in which its ultimate parts are not themselves mechanisms or free agents.*

The addendum in italics is adequate because a machine, as humans create it, must necessarily have finite levels of organization; that is, it must ultimately be composed of simple parts with a specific shape given to a specific substance, which are not mechanisms themselves.

4.3 What Is a Mechanistic Explanation?

The most complicated work on defining an ME was done in the previous item since, for this model of explanation, explaining is about describing the mechanism that produces a specific phenomenon. Now, it is easier to offer a proper definition, with the caveat that most advocates of this kind of explanation would disagree in part with my characterization, for being too restrictive. However, it will be instrumental for my evaluation of ME's relation with possibilist knowledge.

The first influential definition of ME was given by Ernst Nagel (1961). For this kind of explanation, it means showing the causal structures, activities, and interactions of the system's parts that produce or sustain the phenomenon of interest. The explanans, in this case, describes the mechanism, while the *explanandum* is a type of phenomenon, a capacity, and usually not a particular case of its occurrence.

> Mechanistic Explanation: An explanation where the description of the functioning of a mechanism with its parts, structure, and relations, in some environment, is the explanans, and a kind of phenomenon is the explanandum.

We could characterize ME as a second-order explanation. Instead of explaining unique occurrences of a given phenomenon (such as why a rainbow has formed today), it would explain the mechanism that, when formed, makes it possible for any phenomenon of that type to occur (any rainbows). Thus, it provides the necessary support for explaining a concrete event, whether it is explained through causal chains or arguments.

According to Craver (2007), while the causal explanation of, for example, Salmon was etiological, that is, it explained some particular event by its antecedent causes (such as the movement of the bat giving the ball a seven in the pocket), the ME was constitutive or componential. ME explains not a particular event, but a type of phenomenon by describing the type of mechanism that produces it, that is, the description of a set of entities and their activities organized in a certain way.

MEs have been employed, particularly in biology and neuroscience, with fewer examples in psychology. For instance, the regulation of circadian rhythms in *Drosophila* was mechanistically explained through detailed studies of genetic and molecular mechanisms (Rosbash & Hall, 1989). This model shows how specific genes and proteins interact to maintain biological rhythms. I will work here, as a paradigmatic example, with Craver's (2007) description of the contemporary neuroscientific explanation of vesicular neurotransmitter release.

4.3.1 A Paradigmatic Example

Let us then use the example that Craver (2007) presents as a model for ME: the phenomenon of neurotransmitters released by a neuron:

> The mechanism begins, we can say, when an action potential depolarizes the axon terminal and so opens voltage-sensitive calcium (Ca2+) channels in the neuronal membrane. Intracellular Ca2+ concentrations rise, causing more Ca2+ to bind to Ca2+/Calmodulin dependent kinase. The latter phosphorylates synapsin, which frees the transmitter-containing vesicle from the cytoskeleton. At this point, Rab3A and Rab3C target the freed vesicle to release sites in the membrane. Then v-SNARES (such as VAMP), which are incorporated into the vesicle membrane, bind to t-SNARES (such as syntaxin and SNAP-25), which are incorporated into the axon terminal membrane, thereby bringing the vesicle and the membrane next

> to one another. Finally, local influx of Ca2+ at the active zone in the terminal leads this SNARE complex, either acting alone or in concert with other proteins, to open a fusion pore that spans the membrane to the synaptic cleft. (Craver, 2007, pp. 4–5)

To summarize, using examples. ME has in its explanans the specification of a mechanism and in its explanandum a type of phenomenon (in the case above, the vesicular release of neurotransmitters).

The explanans specification always has three components.

1. The entities (parts) of the mechanism have an autonomous location in space and time. In this case, these are N-type Ca^{2+} Channels, Ca^{2+} ions, active zones, a series of intracellular molecules such as Rab3A, Rab3C, VAMP/synaptobrevin, SNAP-25, vesicles containing neurotransmitters, fusion pores, and neural membranes. The criterion derived from Woodward (2003) of mutual manipulability is used to determine which parts are relevant in an ME. A part P is a component of a mechanism if and only if one can change the behavior of the mechanism by intervening in P and change the behavior of P by intervening in the mechanism.
2. The activities of these entities: These activities, which are the productive, causal components of the mechanism, are not mere descriptions of correlations or temporal sequences. They are "legitimated" as part of the explanation when they can be tested via manipulation, expressing causal relationships. In this context, the activities include opening, fixation, diffusion, coupling, fusion, incorporation, phosphorylation, and priming.
3. The organization of these entities: Among the properties that allow entities to participate in various activities are locations, sizes, structures, and orientations. The entities and activities in the mechanisms are organized spatially, temporally, causally, and hierarchically so that, in the example in question, transmitters are released when the axonal terminal depolarizes. The activities of the parts of a mechanism always occur through interaction between parts in a particular temporal and spatial direction, which must be specified in the mechanism. In the example, Rab3A and Rab3C target the released vesicle, not the calcium channels. If the temporal and spatial organization of a mechanism has changed, it is destroyed, or we have another mechanism with another phenomenon as a consequence.

4.3.2 Assessment of Mechanistic Explanations

In "*Explaining the Brain*," Craver (2007, p. 20) tries to offer conditions for assessing MEs. He knows this is fundamental to solidly establish the model, which cannot be merely descriptive (describe what biologists call "mechanistic explanation"); rather, it should be prescriptive (establish the norms to build a proper ME).

However, Craver could not, nor could any other mechanist, offer us anything like the explicit and logical criteria of assessment that the DNM provides. His criteria are based on the general principles of assessing explanations. To Craver, MEs should be judged by (i) the completeness and accuracy in including all components, activities, and organization that cause the phenomenon explained; (ii) the productivity (fruitfulness) in predicting new phenomena based on it; (iii) being consistent with established scientific theories and data and (iv) being simple (parsimony), avoiding any introductions of unnecessary entities or process.

Weiskopf (2011) has a similar prescription (although he does not include productivity and simplicity requirements). To him, the conditions for assessing MEs include (i) Corroboration: MEs should be supported by empirical evidence, which means that scientific data should have confirmed all the components and interactions of a model. (ii) Accuracy: The ME should accurately represent the components and causal interactions of the system. (iii) Power: MEs should genuinely explain the capacity to be explained and not be merely descriptive. (iv) Coherence with knowledge: MEs should not conflict with well-established scientific theories or facts.

Leite (2021) highlights another important distinction for assessing MEs. To him, MEs pass through a continuum that begins with a conjecture about the mechanism underlying the phenomenon and culminates in an actual model that explains how the function is performed in nature. To the mere mechanism-conjecture, he gives the name of "possible mechanism."

It is always relevant to remember that "possible" mechanisms are speculative mechanisms, while "possibilist" mechanisms are those that are necessary conditions for some phenomenon. Therefore, "possible mechanisms" are "only conjectures loosely constructed about the kind of mechanism that could produce the explanandum phenomenon" (Leite, 2021, p. 34). He gives the example of the early work done by simulating cognitive capacities with symbolic digital computers. They could not be taken as actual MEs of cognition, but rather as useful fictions that initiate the exploration of possible ways of performing a function under investigation.

After this, the well-succeeded computational models must be tested in humans and compared with real performances, to begin to constitute what Leite calls "plausible mechanistic explanations," models that match with some empirical evidence about the functioning of the natural mechanism (unlike possible mechanisms) under investigation. Finally, this evolution would culminate in an "actual mechanistic explanation," where the model would describe the exact mechanism in nature, with all its parts, operations, and organization well-known and tested. The boundaries of these levels of evolution, however, seem to me to be excessively subjective.

4.3.3 Mechanistic Explanation and Laws

A specter is haunting mechanicism—the specter of laws.

It seems well-established that Woodward's model significantly improved the usability and intelligibility of the CMEs by demonstrating how factual questions about interventions allow us to identify causes and justify explanations. Through controlled experiments, we can determine which interventions in X effectively change

Y, demonstrating how X causes Y. This, proponents believe, achieves one of the central objectives of CME: eliminating the explanatory role of laws. But does it?

Woodward (2003) believes that the manipulationist model even shows that sometimes laws do not explain anything, exemplified by the classic objection regarding the contraceptive pill for a man (originally by Salmon, 1984). He argues that the law (or generalization) that people who take birth control pills do not get pregnant, coupled with a man taking these pills, does not explain why he does not get pregnant.

Of course, it does not. It is the wrong law.

Explanations are not absolutes but context-dependent, addressing specific questions or doubts. Thus, laws may not always be relevant to certain contexts or questions. They do explain why men do not get pregnant (they do not have a functional female reproductive system: a possibilist law); the problem in Salmon's example was simply that the chosen law was causally irrelevant.

Mechanicists usually confuse laws of nature with scientific laws. Laws of nature regulate the fundamental components of nature and the combinatorial laws arising from their arrangements. Scientific laws are our provisional attempts to represent these natural laws, guiding our indefinite scientific trial-and-error process.

The appeal to manipulation to determine a cause advances the understanding of the nature of scientific explanation. However, just as we manipulate a variable in the experiment to test a law hypothesis, in an explanation, we use the concept of cause, implicitly relying on the concept of law. To say that X causes Y is to say that X always causes Y in the same situation. It is a provisional scientific law. Without this, the statement lacks meaning.

A causal chain, which is the form that a CME takes (A causes B, which causes C, which causes event E), is implicitly based on the idea that an invariable causal relationship, a law, governs each link in this chain. What is being said is not that A causes B, but that for all x, if x is A, then B [$\forall x\,(Ax \rightarrow Bx)$].

Let us analyze the most comfortable example for the CME: the movement of the black ball. It only seems to be explained because we derived it from information about the momentum (mass × velocity) of the white and black balls before the collision of the first with the second, plus the universal law of conservation of linear momentum (assuming a perfectly elastic collision). As this knowledge of the law is assumed, we are satisfied with what is explained in the causal chain of the explanans.

In other words, to be completely explanatory, a causal chain would have to express, in each link of the chain, the DNM that supports that partial explanandum. We do not do this because, many times, the regularity of the relationship is established without knowing the natural law governing it, and other times we do not because stating these laws would be pragmatically redundant for the shared knowledge between the explainer and the inquirer.

When an ME justifies the absence of laws in its formulation by saying that a supposed underlying mechanism can explain each link in a causal chain, it only leads to an infinite regression. Descent to deeper levels of organization of reality would have, at some point, reached the fundamental particles and, therefore, to fundamental laws to explain.

We should recognize that, usually, neomechanicists are not saying that all scientific explanations are explanations of mechanisms, even less that there are no laws at all. They are simply saying that in special sciences, especially biology and neuroscience, we cannot establish general laws to explain the phenomena. I agree, as far as sufficiency laws are concerned, with this. The problem is to make the also usual illegitimate move from the epistemic limitation that we do not find these laws in science to the metaphysical dogma that there are no such laws in nature.

Cummins (2010) expressed the causal explanation tradition's distrust of laws in the DNM, stating that laws "simply tell us what happens; they do not tell us why or how" (p. 284). Explaining the special sciences (which are not fundamental sciences) is not about discovering fundamental natural laws but explaining the regularities (which accomplish the same role as the laws) that govern types of mechanisms (such as the brain or human body systems). Moreover, to explain these regularities is to describe the mechanisms that support them, in the sense that Cartwright (1999) gave to the term "nomological machine."

In conclusion, the explanation process in special sciences involves both regularities and the mechanisms supporting them. However, a comprehensive explanation must ultimately address the fundamental components and natural laws governing these mechanisms. The explanatory role of laws in MEs hinges on how much detail and levels of explanation the explanation-seeker demands.

4.3.4 Evaluation of Mechanistic Explanation

Evaluate this addendum that Craver (2007) makes to his paradigmatic example of ME, the phenomenon of neurotransmitter release:

> the relationship between action potentials and the release of neurotransmitters is stochastic. Only 10–20 percent of action potentials (and instances of Ca2+ influx) eventuate in release events. It is clearly not required of this explanatory relationship that all action potentials eventuate in release events. Nor is it required that no release events occur without action potentials. Nor is it required that action potentials make release events probable, for they evidently do not. (2007, p. 26)

Does this seem explanatory?

No, it is not explanatory in the traditional sense. If we consider a mechanism as something that performs a function, MEs are not fundamentally different from complex deductive-nomological (DN) explanations (DNEs). Both would explain how a specific type of phenomenon becomes necessary given a specific mechanism (acting as a complex law) and particular initial conditions (acting as antecedent conditions of the DNM).

However, if we do not view a mechanism as performing a function, not even probabilistically, then what does it explain?

The challenges of describing a real mechanism as a function are immense, comparable to the challenges of formulating DNEs in the special sciences. The difference lies in the clarity of the logical structure and assessment criteria: DNMs are crystal clear, whereas ME models remain incipient.

I may have been too harsh in my critique and given the wrong impression that I generally dismiss MEs. I do not. I know that MEs offer paramount and irreplaceable insights into how complex systems function, especially in contexts where strict laws may not fully capture the dynamic interactions of components, which, in the case of sufficiency strict laws, happens to be always.

I just think that, first, nothing can take the explanatory role of laws out. Second, MEs are still incomplete in definition and formalization. I believe, with advocates of CME, that explanations must be based on causality. However, I also believe, with advocates of DNE, that they must be ultimately based on laws of nature, at least implicitly. More than that, I believe laws are essential for the metaphysical understanding of causality itself.

An ideally complete explanation in any special science would have to go deeper until the nonreducible arrangements of fundamental particles regulated by laws of nature make higher-level mechanisms work and, successively, until the higher-level mechanism. This is just a regulatory ideal, obviously.

In any case, even if we dismiss DNM, MEs do not encompass all the explanations we require in science. There are scientific explanations that are not causal, such as reductive "theoretical identifications" ("temperature is nothing more than average molecular kinetic energy" or "Water is H_2O") that simply reduce one phenomenon to another. Another type of noncausal scientific explanation is emphasized by the integrative model of explanation (Kitcher, 1989), which argues that scientific explanation aims to offer a unified picture of a vast set of different phenomena. Considering everything so far, I am compelled to adopt a pluralist view of scientific explanation. That is, science sometimes calls for reductive explanations, other times for integrative ones, and sometimes for DNEs or MEs.

4.4 Possibilism and Mechanistic Explanation

When used to explain the occurrence of a particular case of a phenomenon, an ME aims to be a sufficiency explanation, either explicitly or implicitly. However, this is rarely the case. More often, MEs explain the mechanism itself, which is inherently a PE.

Let me explain. Consider an ME of the rainbow I see from my window today. Such an explanation typically describes the mechanism that enables the formation of a rainbow. For the ME to fully explain today's particular event, it would need to account not only for the mechanism of rainbow formation but also for the specific conditions that activated the mechanism today.

In other words, the description of a mechanism constitutes an enormous and complex necessary but not sufficient condition for the occurrence of a phenomenon. To be both necessary and sufficient, an ME must also include the occurrence of the proper input. Therefore, describing the presence of a specific mechanism or condition that supports a particular phenomenon, which initially seemed impossible or unlikely, is a form of a PE.

During my study of possibilism, a common question from colleagues and reviewers was: "Your framework seems to fit well within causal explanations, so why present it in a covering law model?" It is time to address this question here.

PEs can explain the possibility or impossibility of events either through a deductive argument or as a causal chain. However, PEs in the form of a causal chain are informatively poorer because they lack the explicitation of the possibilist law regulating the event. In some cases, such as Dray's baseball catching example, the relevant possibilist law (if the ball is not within reach, it cannot be caught) is trivial and may not seem necessary. In other cases, such as the understanding of combustion, humanity has only recently become aware of the underlying possibilist law.

Furthermore, CMEs are formally less robust than DNEs. Possibilist laws and explanations, being demonstrable through formal and DN means, adapt more easily to models of causal explanation. The reverse is not true; causal models cannot easily accommodate the formal rigor of PEs.

4.4.1 Possibilist Causal Explanation

When considering the causal explanation of behavior or any case study in the human sciences, we can employ the causal approach of the necessary condition proposed by Levy and Goertz (2007). In this framework, an outcome occurs only if all required conditions are in their place in the chain of events; the absence of any one necessary condition would make the outcome impossible:

> In a chain consisting of necessary conditions, if any link is absent then, counterfactually, the outcome would not have occurred. If an historical narrative describes a causal chain, then we would have a temporal series of necessary conditions for the event in question. (2007, p. 24)

These chains are temporal: each element must be present in a particular order for the outcome to materialize, though no one alone is sufficient. For example, in explaining the outbreak of the First World War, Levy and Goertz (2007) identify key events—such as Germany's assumption of British neutrality, its support for Austria-Hungary, and Russia's mobilization decisions—as necessary conditions for it. Each of these events, although not sufficient, was indispensable. If any of it had been absent, the war involving all major powers would not have occurred.

Levy and Goertz (2007) stress that necessary condition causal explanations are incomplete, while sufficient causal explanations are not: "The existence of a necessary condition makes the next link possible, but usually other factors must be included to explain why it in fact did happen" (p. 26).

This shows that adapting my deductive PE model to a causal PE model does not go smoothly. Let us look at a PE in the form of a causal chain by resorting again to Dray's (1954) classic example, which he took from a Canadian variety magazine: "At the back of center field was a high platform for the scoreboard operator. The central defender climbed the ladder and caught the ball six meters from the ground" (p. 18). We could reconstruct the explanation as follows: The player ran to a staircase, climbed to a platform four meters above the ground, stood in the trajectory of the ball, and, with his arm extended, made it possible to catch it.

What is different about this form of causal chain from the "ball seven" CME example? It is that at various moments in this chain of events we have an agent's decision that is either not determined (not necessary) or, if it is, we will probably never be able to manipulate or predict completely. The player's decision to run to the stairs is the initiator of the causal chain here, and, at the end of it, the catching of the ball was not necessarily caused; what was caused was only its possibility. That is so because the event, possible at that time, still depended on the player's desire and performance in that play to happen.

The presence of free agents (or at least agents whose ultimate cause of their behavior is not known to us) in links in a causal chain makes any causal explanation a mere explanation of the epistemic possibility of an event since their decision was not (or we do not know why it was) necessary.

4.4.2 Possibilist Mechanistic Explanation

To begin discussing this topic, it is better to remember a distinction already made here. Neither Possibilist nor Dray's How-Possibly Explanations are Possible Explanations. However, when authors like Craver (2007) or Glennan (2017) talk about how-possibly MEs, they refer to "unrestricted conjectures about the type of mechanism that might be sufficient to produce the phenomenon of the explanandum" (Craver, 2007, p. 112). These are mere hypotheses about how a set of parts and activities might be organized to produce the phenomenon to be explained, even if we do not know whether the conjectured parts exist, have activities, or are actually organized as in the model. In other words, they discuss possible (not possibilist) explanations.

This reinforces my judgment that the term "how-possibly" has been degraded and is used today in the philosophy of science with a different meaning from the original.

Suppose we accept the definition of a mechanism as an executor of functions. In that case, explaining a mechanism means explaining how a particular type of phenomenon becomes necessary in the presence of a specific mechanism (acting as a large and complex law) and specific initial stimuli and environments (acting as the antecedent conditions of the DNM).

However, when mechanisms are used to explain agents' behaviors, they function variously as permissive conditions, necessary conditions, or contributors to the occurrence of an event.

Let us consider the most typical fields of MEs: psychology and neuroscience. Psychological explanations largely explain human capabilities (Fodor, 1991a; Wright & Bechtel, 2007; Craver, 2007), which are considered effects of mechanisms. We offer a PE when we use mechanisms or their parts to explain why a particular behavior is possible or impossible.

Take, for example, the mechanism for forming episodic long-term memories. If the explanandum is the phenomenon of memory formation, the hippocampus is just a component of the mechanism that explains it. However, if we use this model to explain how Jonas remembers a book from his childhood, the most we can do is show that forming this memory was possible with intact hippocampi.

This limitation does not arise only because Jonas could have decided never to think about the book again, thus preventing the repetition fundamental to forming these memories. It arises because each non-redundant part[2] of a mechanism becomes a necessary condition for its correct functioning and expresses a possibilist law (e.g., if the hippocampi are completely severed, it is not possible to form new explicit long-term memories).

In other words, as a mechanism is a stable causal chain, if it lacks a redundant path to perform a function, that causal link[3] is a necessary (not sufficient) condition for the explanandum. Break that link, and the chain is broken; the regular effect is no longer possible. While the mechanism as a whole functions as a complex possibilist law, each non-redundant part functions as a simple possibilist law: if any such part is missing, the effect (explanandum) becomes impossible.

Let us return to Craver's (2007) example of an ME for a neurotransmitter release by a neuron. Remember that the mechanism is initially activated by an action potential that depolarizes the axonal terminal, thus opening the calcium channels (Ca^{2+}). It is almost impossible, therefore, for us to explain the particular event of neurotransmitter release by a single neuron except in artificial laboratory conditions. However, any link in the long chain of entities and activities of this mechanism becomes a necessary condition for the release of neurotransmitters by a neuron.

Moreover, suppose our interest is, for example, to prevent this release. In that case, we can work on psychotropic drugs that prevent the binding of Ca^{2+} to the kinase, the phosphorization of synapsin, or the binding of v-SNARES to t-SNARES, among other options. The explanation of why a particular neuron did not release neurotransmitters in response to one of these drugs will be possibilist, showing that, for example, synapsin, which was a permissive condition for the release of the transmitter-containing vesicle from the cytoskeleton, was inhibited. As this vesicle was a necessary condition for releasing neurotransmitters, this release became impossible.

4.4.3 Possibilism and Negative Causality

This leads us to another issue that Craver (2007) addressed, a widely debated concept in the metaphysics of causality and philosophy of science: the concept of negative causality. More precisely, the prevention and omission activities are commonly described in mechanisms.

The controversial concept of "negative causation" refers to cases where the absence of something, rather than its presence, "causes" or is "caused" by an event. It is linked with debates about whether non-events (such as not doing or not happening) can be considered genuine causes. Negative causation has two primary forms: prevention and omission.

What we call "prevention" occurs when an action or event precludes another event from happening. Here, an intervention changes the course of events, which is consistent with the concept of cause. The puzzle is that the supposed consequence is the non-occurrence of something, and this precluded event was an expected or possible outcome. For instance, taking an antibiotic can prevent an infection.

When it comes to "omission," the problem is even worse. Because it occurs when something does not act or behave in such a way that can prevent an event, it leads to a particular outcome. Involving mechanisms, one example that can be given is when a safety system that prevents a nuclear reactor from overheating fails, "causing" its meltdown. Therefore, in the case of omissions, we would have a non-occurrence that causes something.

There is no such thing as nothing causing something or something causing nothing. When we talk of "prevention" (something causing nothing) in a mechanism, we strictly refer to a stopped process. Something has caused these interruptions of movement and changes in course. Taking an antibiotic inhibits protein synthesis and DNA replication in bacteria, killing them or blocking their growth, stopping the infection process already in progress.

Let us return to the paradigmatic example of Craver (2007) of the Long-Term Potentiation (LTP) mechanism. Mg^{2+} ions block the NMDA receptor channel and prevent Ca^{2+} from flowing into the postsynaptic neuron. The Mg^{2+} block must be removed to induce LTP. Depolarization of the postsynaptic neuron causes Ca^{2+} to enter through the NMDA receptor (Craver, 2007, p. 80).

4.4.3.1 Omissions Are No Causes

On the other hand, when we talk of "omission" (nothing causing something), we are referring to a process already in course, caused by another thing, that something could stop but did not. And it did not because of other causal processes that had "prevented" it. Then, again, something caused something.

For example, in the Fukushima nuclear disaster, the cooling systems designed to cool the reactor's nucleus did not act because they were damaged by the flood, as were the other diesel generators that provided backup power for the spare system.

We should keep in mind that we cannot say that the absence of Mg^{2+} causes the entry of Ca^{2+} into the cell because an absence is not a process; it is not a thing, a component, or an activity. And absences, in other words, nothingness, have no causal powers.

Moreover, another problem with considering omissions as causes of events is that they are virtually infinite. Let us consider that the absence of Mg^{2+} causes the entry of Ca^{2+} into the cell. We also must consider that the absence of incineration of the cell causes it, the absence of any other ionized substance also causes it, and so on.

By comparison, I could be accused of causing all the murders that have occurred since I was born by not being there to prevent them or of causing the existence of the Mona Lisa by not destroying it when I visited the Louvre (or when, instead of visiting the Louvre, I decided to write this book).

4.4.3.2 Preventive Conditions

I do not see how to consider omission cases as legitimate cases of causality, but their counterpart reveals the apparent mystery of their explanatory significance: prevention process. Preventions are the processes by which a given cause inhibits or prevents an effect. The strangeness with the statement that "the absence of Mg^{2+} causes the entry of Ca^{2+} into the cell" disappears when we reformulate it to state that "the presence of Mg^{2+} causes the non-entry of Ca^{2+} into the postsynaptic neuron."

It is wrong to call prevention "negative causality." The preventer is a presence; it is something. Therefore, it has causal powers. Furthermore, although there may be many distinct preventers for the same effect, whether V prevents or not E is a matter of fact rather than possibility; in other words, we can multiply indefinitely the things that by omission "caused" E, but if something prevented E from occurring, then that something is unique.

Prevention activities indicate the presence of possibilist laws. When we state that the presence of Mg^{2+} causes the non-entry of Ca^{2+} into the postsynaptic neuron, we are faced with the statement that Mg^{2+} eliminates a necessary condition for the entry of Ca^{2+} into the neuron (such as a certain pattern of ionization). In this case, we could call Mg^{2+} the preventive condition, the negative counterpart of the permissive condition. What a preventive condition eliminates is a necessary condition for a given event, not the event. In other words, the preventive condition, by eliminating a necessary condition for E, eliminates the possibility of E.

Let us review. Given that N is a necessary condition to E ($\neg N \rightarrow \neg E$), a permissive condition P for an event E in a situation S is a sufficient but not necessary condition for a necessary condition N ($P \rightarrow N$). In other words, a permissive condition for an event does not imply the event E itself but, rather, a necessary condition for its occurrence.

In its turn, a preventive condition V to E is a sufficient but not necessary condition for the absence of a necessary condition N ($V \rightarrow \neg N$) to E. In other words, a preventive condition for an event precludes it by blocking a necessary condition for it.

Let us analyze it.

Why should we not directly consider a preventive condition a sufficient condition for the non-occurrence of E ($V \rightarrow \neg E$)? Because, as we already saw (Cartwright, 1983), there are no such purely logical relations in the physical world, what there is, are causal pressures,[4] sometimes, *almost* absolute. Mumford and Anjum (2011) interpret preventers as such causal pressures: "Two powers P1 and P2 that dispose in opposite directions with equal magnitude can be each other's preventers, if we get an equilibrium state, for instance" (p. 165).

However, I would like to use the term "prevention" more specifically to distinguish two different causal processes. In Mumford and Anjum's case, we would be talking about negative contributory conditions, not preventers.

Imagine a woman who periodically experiences painful herpes outbreaks. She must have the necessary condition: the varicella-zoster virus in her body. She seeks to prevent future outbreaks and manage her condition. Hypothetically, she has two different causal processes to consider: fighting the action of the virus within her organism (the symptom) or eliminating the virus from her organism (the necessary condition). Nowadays, we only have the technology to help the first fight.

To combat the virus's action, she can take an antiviral medication that is a contributory condition against herpes zoster eruptions by slowing down the replication of the varicella-zoster virus. A (real) example is the medication valacyclovir, which, in the body, is converted to acyclovir, which blocks viral DNA polymerase, which is needed for the reproduction of the virus. This reduces the frequency and severity of outbreaks, promoting quicker healing during active episodes. Another contributory condition is reinforcing her immune system by taking supplements like vitamin C or *Echinacea*, which will also exert causal pressure against the eruptions. *Contributory conditions* exert causal pressure directly against (or pro) an effect.

On the other hand, a *preventive condition* exerts overwhelming causal pressure against a necessary condition for this effect. It would be a new drug that completely eradicates the varicella-zoster virus from the body. If valacyclovir would act by killing the existent virus, it would be a preventive condition to eruptions. Still, as it only acts to lower the frequency of the virus multiplying, it is only a contributory condition[5] (it does not act to eliminate the necessary condition but to diminish the effect of the proliferation of the virus). A hypothetical drug that would be an actual preventive condition would target and destroy the virus at a molecular level, abolishing its possibility. It would prevent it from causing future outbreaks. Another hypothetical preventive condition for herpes eruptions would be a drug that cracks some necessary (non-redundant) chain in the mechanism through which the virus provokes the eruptions. Nowadays, however, the only preventive condition to prevent herpes eruptions is not having physical contact with carriers of the virus.

In summary, in managing herpes zoster eruptions, valacyclovir (and Vitamin C and *Echinacea*) acts as a contributory condition by inhibiting viral replication, while a hypothetical drug that eradicates the varicella-zoster virus would be a true preventive condition. The former diminishes the effect; the latter eliminates the necessary condition for the effect.

This distinction is crucial: a contributory condition (like valacyclovir) exerts causal pressure against an effect, while a preventive condition (hypothetical antiviral drug) precludes the effect by targeting its necessary conditions.

When we know a preventive condition V, we already know *that* V prevents E. What we may still need to know is *how*. Event E does not exist yet. How can we eliminate something that does not exist yet? *By eliminating its possibility*. Some necessary conditions for E already exist in a mechanism. Therefore, a V can cause its elimination.

We should ask ourselves why condition V precluded E in a mechanism or situation where E was expected. In a mechanism operating within an environment (situation S), what we can inadvertently call a sufficient condition for E is simply, at its best, one more necessary condition that, conjointly with all the others, forms a necessary and sufficient condition for E.

On the other hand, why should we not consider the preventive condition a sufficient condition for the non-occurrence of a sufficient condition for E ($V \rightarrow C \rightarrow \neg E$)? Because if it were so, on several occasions, even in the presence of this supposed preventive condition, E would happen, caused by other sufficient conditions for it (we block the gun, and the assassin kills with a knife or a stone). That is why we should prevent E (the killing) by preventing a necessary condition to it (taking the target from the range of the assassin). This is the best way for something to be a preventer: not by eliminating an event that does not exist yet, but by eliminating a necessary condition that already exists.

Furthermore, what is an omission, after all? An omission is the absence of a preventive condition V for E. Nothing more.

Therefore, a preventive condition for E:

a) disables E if it is in a particular context S (Mg^{2+} ions (V) prevent Ca^{2+} from flowing into the postsynaptic neuron (E) in neuronal neurotransmitters mechanism (S));
b) disables E by eliminating a necessary condition to it in S (Mg^{2+} ions (V) block the NMDA receptor channel (N) and prevent Ca^{2+} from flowing into the postsynaptic neuron (E));
c) is an event among others (V1, V2, V3,…) that makes another event E impossible (by also blocking the NMDA receptor channel or any other necessary condition to E).

Prevention and omission are, therefore, just ways of talking, words that we use to refer to more complex configurations of conditions and causes. They are concepts imported from ethics, where they make every sense because they refer to acts from free and conscious moral agents. In ethics, prevention and omission are fundamental to moral and practical problems of responsibility, duty, the assessment of liability in various situations, and the consequences of actions or inactions. For example, suppose the only doctor on duty decides not to treat a patient admitted to the emergency room with extensive bleeding. In that case, the omission (not acting) will be causally linked to the outcome (the patient's death). On the other hand, stopping a child from going to a dangerous and open platform will be causally linked to the outcome (the child does not die).

These concepts make so much sense in ethics because there, they are applied to free and conscious agents, who would decide to act to prevent an event on the course or decide freely not to act and, by not using their originary and free causal powers, let something happen. But this is another story. We are not talking here about ethics; instead, we are investigating mechanisms. And mechanisms have not free agents within their parts. On these, prevention and omission are only ways of talking, descriptive shortcuts that are directly linked with possibilist knowledge.

This entire digression here was necessary to clarify that when we explain why a mechanism, in the presence of adequate input, did not work due to some preventer, we should offer a PE. To account for its failure, given that proper functioning was expected, we must explain why the unexpected event itself was possible.

Concluding the subitem, it seems clear that PEs can take a causal format. In fact, a PE does not need to explicitly rely on laws. Nevertheless, it is more informative with them. In other words, they are, in some contexts, better explanations.

4.5 Final Remarks

This chapter introduced the rise and application of MEs within science, primarily in biology. It has defended the peculiarity of contemporary interpretations, which present mechanistic models as complex, probabilistic, and somehow confused phenomena. Departing from the simplistic, deterministic views of early mechanicism, neomechanicism maintains its compromise with determinism as a metaphysical position.

Various definitions of "mechanism" proposed by leading figures such as Cartwright, Glennan, and Craver were critically assessed. It was argued that they are still vague and overly broad, conflating mechanisms with any form of causal interaction. To address this, I have presented a more precise definition of "mechanism" as a (mathematical) function.

While MEs have been recognized as invaluable in fields such as biology, neuroscience, and psychology, they have yet to be supported by sufficient empirical case studies to demonstrate their practical applicability. They often face challenges in addressing real phenomena in complex systems, and today, they are in no better state than DNM.

This chapter also highlighted the vital role of MEs in providing necessary conditions for phenomena, which make them truly PEs. In recognizing MEs as forms of possibilist knowledge, we find that both approaches are complementary.

MEs, focusing on causal interactions and system functions, provide another framework for exploring how specific conditions can make certain phenomena possible or impossible. This alignment highlights the importance of a pluralistic approach to scientific explanation, where laws and mechanisms contribute to a comprehensive understanding of the natural world.

I have already stated and reaffirmed here that I share with logical positivism the belief that scientific explanations must, at the very least, implicitly contain laws. However, I want to emphasize again that the ideas defended here are not committed to the logical positivist denial of the concept of cause. There is nothing to prevent the use of permissive, preventive, and necessary conditions in an ME.

I do not believe that Cartwright, Woodward, Craver, or the entire causal-mechanical approach is committed to banning laws or generalizations from scientific explanations, especially the kind of laws, possibilist ones, I have presented here. However, this discussion is not the central objective of this book. The central objective of this book is something much more difficult: demonstrating that possibilist knowledge is clear enough to be formalized into deductive explanations and predictive laws. And this, I believe, has already been achieved.

5

Sciences and Possibilism

This chapter presents examples of possibilist laws and explanations, beginning with reflections that help identify possibilist laws in science. Following this, several examples of possibilist laws in physics are discussed, as well as from biology, history (where William Dray first presented how-possibly explanations [HPEs]), geography (which has a possibilist theory), sociology (as presented in Goertz's book *Necessary Conditions*), and economics. Especially in history and geography, the emphasis will be on examples similar to possibilist explanations, as they are derived from the works of Dray and La Blache. However, they will be able to reinforce my thesis that even idiographic disciplines use possibilist reasoning. The predictive aspect of possibilism is emphasized in physics and biology. As for the examples in social sciences, the explanatory and predictive roles of this approach are more balanced.

5.1 Trivial Laws? Not That Much

Possibilist laws are everywhere. Simply because necessary conditions are everywhere. Practically every law of any domain, even physics, can generate them. That means that many seemingly sufficient laws can generate possibilist laws, but not in the way that may appear at first sight.

First, it is evident to any primary student of logic that a sufficient condition to some result (shooting in the head to kill an animal—$S \rightarrow K$) is not necessarily a necessary condition to it (there is no need to be shot in the head for an animal to die). In other words, none can negate the possibility of a result K by negating one of its sufficient conditions S ($\neg S \rightarrow \neg K$). It is an instance of the basic "denying the antecedent" fallacy.

In another way, it is expressed with the formulas used here. Given the sufficiency law $\forall x(C(x) \rightarrow R(x))$. If we try to infer from it a possibilist law using the same condition—as in $\forall x(\neg C(x) \rightarrow \neg \Diamond R(x))$—it would simply be invalid because while C(x) guarantees R(x), the absence of C(x) does not rule out R(x) happening by other sufficient causes (the animal can be stabbed in the heart, burned, drowned, and so on).

Despite this, sometimes this invalid operation can yield a formula that is adequate to reality if the sufficient condition is also a necessary condition for the consequent.

Why have I said that every sufficiency law can generate possibility laws? They will not always appear in the formula or the equation of the law, but they always will be hidden in the ceteris paribus clauses. As we have seen, there are no such things as sufficiency strict laws outside the fundamental laws of reality (laws of nature governing fundamental forces). Even in physics, what we call "scientific laws" are simplified idealizations for the sake of calculation, which requires the clause that all other influencing factors are held constant (Cartwright, 1983). Among them, there are always necessary conditions for the phenomenon.

This problem brings us another: the falsifiability of every scientific law. Just as sufficiency laws are frequently falsified, momentary corroborated scientific possibilist laws can be falsified when we discover that the result has occurred in the absence of the supposed necessary condition. That is not a defect of possibilist laws; on the contrary, this is precisely why they can be scientific. Usually, when this happens, it is because we, at the present moment, are taking permissive conditions as necessary conditions for something to occur. The falsification of such a false possibilist law is an opportunity to deepen our understanding of reality, uncovering an actual necessary condition (or one-level-down permissive condition) for something on a deeper level than the falsified candidate.

Before we embark on our quest to uncover possibilist laws across the sciences, let us consider two crucial caveats:

First, the significant number of examples of laws in all major disciplines that we are about to evaluate are not currently presented as possibilist laws in science. However, what I am about to demonstrate is that they could be. The possibilist laws are there, even if we are not currently seeing them.

Second, as we have already seen, not every transformation in a law expressed by an equation makes causal sense. This is a crucial point to understand, as equations do not always express the necessary precedence in time from conditions that are causes of other variables. To illustrate this problem, the examples of possibilist laws in physics will start with a very simple equation, Ohm's Law.

Nevertheless, first, let us summarize the five directions to look for (and find) possibilist laws in science:

1) A sufficient condition cannot automatically turn into a necessary condition.
2) There are often necessary conditions hidden in the *ceteris paribus* clauses.
3) Today's possibilist law can be falsified tomorrow; it happens.
4) There are often necessary conditions hidden in Mackie's "disjunction of conjunctions" (review item 2.5.5).
5) Not all the variables on the complex side of an equation can be a necessary condition for the simple side, only the causally relevant ones can be.

Now, let's go hunting.

5.2 Possibilism in Physics

Here, first, I will use Ohm's Law to illustrate the problem of causal relevance in equations to the derivation of possibilist laws. Second, I will show how possibilist laws can make simpler some puzzling laws of quantum physics. Finally, I will offer examples of impossibilist physical laws.

5.2.1 Ohm's Law and the Problem of Equations

Ohm's Law postulates that the voltage across a conductive circuit is directly proportional to the current flowing through it, given that the resistance remains constant. It is usually represented as $V=I \times R$, where V is the voltage across the circuit, measured in volts (V), I is the current in amperes (A), and R is the resistance of the circuit and its components in ohms (Ω).

Ohm's Law can be interpreted as implying a possibilist constraint, expressible as a possibilist law with the form $\forall x\,(\neg Vx \rightarrow \neg\Diamond Ix)$, which says something like "for any electrical conductor, if there is no voltage difference applied across it, then an electric current cannot flow through it." So, in the used formula, Vx stands for "a voltage difference is applied across conductor x," and Ix stands for "an electric current flows through conductor x."

We could theoretically derive other possibilist laws from this equation because none of its terms is a constant. For example, we could think of a PL that said the reverse: $\forall x\,(\neg Ix \rightarrow \neg\Diamond Vx)$, which means "if there is no electric current, then no voltage is possible." This formulation is logically derivable, but not sound because $V \rightarrow I$ is not itself a strict law. It would need to hold in nature if the law was true and laws were a logical relation. However, in the real world of causality, we understand that voltage causes the current to flow through a circuit with a specific resistance. In practical scenarios, such as in high school electronic labs, we use a voltage source—a battery—to create a potential difference between points in the circuit to generate current.

Thus, while the reverse possibilist law $\forall x(\neg Ix \rightarrow \neg\Diamond Vx)$ is logically correct, it does not hold in physical reality, since voltage can exist without current—for example, in an open circuit where Ohm's Law does not apply due to the violation of its ceteris paribus conditions.

In summary, if we aim to maintain possibilist laws as causally meaningful, we should observe that not every logically valid derivation from an equation will correspond to physical reality.

5.2.2 Possibilist Laws in Quantum Physics

How does possibilism perform in quantum physics? The field of the infinitesimal is supposed to have an inherently probabilistic nature. Despite my limited knowledge of the area, it seems to be a rich field for possibilist laws. Although I wish to rigorously test my metaphysical speculations, my mathematical skills are limited. Nevertheless, it is easy to show how some possibilist laws are derived from the field's most well-known principles.

5.2.2.1 *Heisenberg's Uncertainty Principle*

Perhaps the most famous result of quantum mechanics is the *uncertainty principle*. Roughly, it states that it is impossible to precisely determine a quantum particle's position and momentum simultaneously (Nielsen & Chuang, 2011, p. 89). From this, we can extract the following impossibilist law: *For any quantum particle, if the position is precisely determined, then it is impossible to precisely determine its momentum*, and vice versa.

$$\forall x\,(Px \rightarrow \neg\Diamond Mx)$$

where
 Px: "the Position of particle x is precisely determined."
 Mx: "the Momentum of particle x is precisely determined."

5.2.2.2 *Superposition Principle*

In quantum mechanics, it is assumed that particles can exist in multiple states simultaneously until they are measured (Nielsen & Chuang, 2011, p. 94). From the concept of superposition, we can deduce the following possibilist law: For any quantum system, if its wave function has collapsed due to measurement, then superposition is not possible (it cannot exhibit behavior resulting from being in multiple states simultaneously).

$$\forall x\,(\neg Mx \rightarrow \neg\Diamond Sx)$$

where
 Mx: "system x has been Measured."
 Sx: "system x is in a Superposition of states."

Some enlightenment. According to quantum theory, every quantum system has a wave function that describes its state. However, when its state is measured, the wave function "collapses" to one of its possible states. This wave function, therefore, does not vanish or cease to exist; it only changes.

5.2.2.3 *No-Cloning Theorem*

This principle is already a possibilist law. Wootters and Zurek (1982) stated that creating a precise copy of an arbitrary unknown quantum state is impossible. Restating it to assume the traditional form of a possibilist law, we have: *For any quantum state, if it is not known and ordered, then it is not possible to exactly clone it.*

$$\forall x\,(\neg Kx \rightarrow \neg\Diamond Cx)$$

where
 Kx: "state x is orthogonal, Known, and ordered."
 Cx: "state x exactly Cloned."

Nielsen and Chuang (2011, p. 532) concluded from this possibilist law that, according to our current scientific knowledge, the cloning of non-orthogonal pure states[1] remains impossible "unless one is willing to tolerate a finite loss of fidelity in the copied states." A similar conclusion is extended to mixed states. This means that exact teleportation is impossible. It is a matter of possibilist thinking in deciding not to test such a sci-fi device.

These examples of possibilist laws in quantum physics illustrate how the fundamental nature of physics relies on possibilities and impossibilities, reflecting the underlying non-classical, counterintuitive, and indeterminist nature of quantum phenomena, which are likely totally governed by laws. In Chapter 7, I will dare again to speculate a little more on the role of possibilism in the fundamental realm of reality.

5.2.3 Possibilist and Impossibilist Laws in General Physics

Let us examine some possibilist laws in other subfields of physics, as well as related fields such as astronomy and chemistry. But first, three examples of physical impossibilist laws.

5.2.3.1 Traveling Faster Than Light

Einstein's principle is the iconic impossibilist law. It states that for any particle with mass, it is impossible for it to travel faster than the speed of light.

$$\forall x\ (Mx \rightarrow \neg \Diamond Fx)$$

where

Mx stands for "the particle x has Mass."

Fx stands for "the particle x travels Faster than the speed of light."

5.2.3.2 Reaching Absolute Zero Temperature

Another widely known impossibilist law is the "absolute zero Kelvin." It states that no physical system can reach absolute zero temperature.

$$\forall x\ (Px \rightarrow \neg \Diamond Zx)$$

where

Px stands for "the system x is a Physical system."

Zx stands for "the system x reaches absolute Zero temperature."

5.2.3.3 Perpetual Motion

The problem of perpetual motion is an impossibilist law related to the problem of energy production. It states that, for any machine, it is impossible to produce more energy than it consumes.

$$\forall x\,(Mx \rightarrow \neg\Diamond Ex)$$

where
Mx stands for "the entity x is a machine."
Ex stands for "the machine x produces more Energy than it consumes."

5.2.3.4 *Conservation of Energy*

This possibilist law states that for any closed system, if no energy enters or leaves the system, then the total energy of the system cannot change.

$$\forall x\,(\neg Ex \rightarrow \neg\Diamond Cx)$$

where
Ex goes for "Energy enters or leaves the system."
Cx for "total energy of the system Changes."

5.2.3.5 *Formation of Ionic Bonds*

Another possibilist law posits that for any pair of atoms, if there is no transfer of electrons, then an ionic bond cannot form between them.

$$\forall x\,(\neg Tx \rightarrow \neg\Diamond Ix)$$

where
Tx could be "electrons are transferred between atoms in x."
Ix could be "an ionic bond forms in x."

5.2.3.6 *Visibility of Eclipses*

This simple possibilist law can be derived from prosaic facts of astronomy. It states that, for any location on Earth, if the Sun, Earth, and Moon are not aligned, then it is not possible to observe a solar or lunar eclipse from this specific location.

$$\forall x\,(\neg Ax \rightarrow \neg\Diamond Ex)$$

where
Ax could be "Sun, Earth, and Moon are aligned relative to location x."
Ex could be "an eclipse is visible from location x."

5.2.3.7 *Law of Inertia*

Another iconic possibilist law, it states that for any object, if no external force is applied to it, then it is impossible for the object to change its state of motion.

$$\forall x\,(\neg Fx \rightarrow \neg\Diamond Vx)$$

where
Fx could be "a net external force is applied to object x."
Vx could be "object x changes its velocity."

5.2.3.8 First Law of Thermodynamics

A last possibilist law: if there is no temperature difference between two different bodies, then heat cannot flow between them.

$$\forall x\ (\neg Tx \rightarrow \neg\Diamond Hx)$$

where
Tx could be "there is a temperature difference between body x and its surroundings."
Hx could be "heat flows from body x to its surroundings or vice versa."

5.3 Possibilism in Biology

Regular high school textbooks are full of biological laws that can be read as possibilist. I will spare the reader any further examples from genetics and triggers, as well as viruses and bacteria and their diseases. I think I made clear that all diseases that depend on some gene to happen generate a possibilist law of type $\forall x\ (\neg Gx \rightarrow \neg\Diamond Dx)$, where Gx could be "a person x has the gene G," and Dx "a person x has the disease D." If a person has not the gene G, it is impossible for her to develop the associated disease D. The same way all the virotic and bacteriological diseases are regulated by a possibilist law as $\forall x\ (\neg Bx \rightarrow \neg\Diamond Dx)$, where Bx could be "a person x is infected by the bacteria B" and Dx "a person x has developed the disease D." We can read the formula as "If a person was not infected by the Bacteria B, then it is impossible for him to develop the associated disease D." Therefore, let us look for different examples of possibilist laws from biology.

5.3.1 Possibilist Laws in Botany

Let us remember some botany laws that we learned in high school, from which we can derive possibilist laws that can teach us how to avoid some botanical events.

5.3.1.1 Germination

For any viable seed, if it does not receive water, then it is not possible for the germination process to occur.

$$\forall x\ (\neg Wx \rightarrow \neg\Diamond Gx)$$

where
Wx is "seed x receives water."
Gx is "seed x germinates."

5.3.1.2 *Photosynthesis*

For any plant, if there is no light available, then it is not possible for photosynthesis to occur.

$$\forall x\ (\neg Lx \rightarrow \neg \Diamond Px)$$

where
Lx is "light is available to plant x."
Px is "plant x performs photosynthesis."

5.3.1.3 *Light for Flowering*

For any plant that requires photoperiodic cues for flowering (to "know" which period of the year it is), if it does not receive the minimal light conditions (specifically the right photoperiod), then it is not possible for it to flower.

$$\forall x\ (\neg Lx \rightarrow \neg \Diamond Fx)$$

where
Lx is "plant x receives the minimum photoperiod for flowering."
Fx is "plant x flowers."

5.3.2 Possibilist Laws in Physiology

These are the building blocks of neuroscience, but we will investigate neuroscientific laws in the Psychology chapter. Let us remember some very well-known textbook laws, now in its possibilist character:

5.3.2.1 *Protein Synthesis*

According to Alberts et al. (2014), for any cell, if no mRNA is available, then it is not possible for protein synthesis to occur.

$$\forall x\ (\neg Mx \rightarrow \neg \Diamond Px)$$

where
Mx is "mRNA is available in cell x."
Px is "protein synthesis occurs in cell x."

5.3.2.2 *Nerve Conduction*

For any neuron, if there is no action potential generated, then it is not possible for nerve impulse conduction to occur.

$$\forall x\ (\neg Ax \rightarrow \neg \Diamond Nx)$$

where
Ax is "action potential is generated in neuron x."
Nx is "nerve impulse is conducted by neuron x."

5.3.2.3 Cellular Respiration

For any organism, if there is no oxygen available, then it is not possible for aerobic respiration to occur.

$\forall x\ (\neg Ox \rightarrow \neg \Diamond Rx)$

where
Ox is "oxygen is available to organism x."
Rx is "organism x performs aerobic respiration."

5.4 Possibilism in History

It makes little sense to talk about laws in history. If there are any laws of the historical process, they are sociological, geological, psychological, and economic, not historical. And they are probably possibilist ones. History is an idiographic discipline; it studies unique and unrepeatable events, individuals, and periods. This kind of approach to knowledge contrasts with the nomothetical (as in physics, chemistry, and biology), which seeks to establish general laws and theories applicable to the same kind of events. On the contrary, history is retrospective (it does not offer predictions) and descriptive (it does not perform experiments).

However, in their primary and constant work of describing and interpreting the past, historians must always offer explanations. That was one of the factors that led Dray, a philosopher of history, to develop the first version of HPEs. Therefore, let us examine some of the HPEs Dray offered in his work, "Laws and Explanation in History," from 1957. It is worth noting that Dray was against laws in explanations. However, if we were to offer historical possibilist explanations here, they would have to be supported by psychological, economic, and sociological laws that apply to historical events.

5.4.1 Dray's Examples of Historical How-Possibly Explanations

Contextualizing, the HPEs are causal and informal explanations with a similar objective as the possibilists. In historical writing, they are instrumental in cases where the historian must explain how a specific historical outcome came about despite earlier conditions suggesting a contrary result. Let us see summaries of Dray's examples (1957).

5.4.1.1 Hanoverian Succession

About the Hanoverian Succession and the Relative Power of the British Crown versus Parliament in 1714, Dray (1957, p. 162) discussed how, given the historical context of monarchs like Elizabeth I, who exerted substantial control over Parliament, it might seem unexpected that the Crown's power would be less than that of Parliament by 1714. The HPE explores how the personal qualities of the new king, the economic powers of Parliament members, and the political climate of the time made it possible for Parliament to gain the upper hand. This shift in power dynamics could be seen as possible due to these contributing factors, which do not necessarily make the outcome inevitable but show it was not entirely unexpected.

5.4.1.2 *The 1640 Dissolution of the Parliament*

Dray here (p. 163) gives an example taken from Ashley regarding the dissolution of the Short Parliament, focusing on the unexpected absence of influential figures like Laud and Strafford at a crucial Privy Council meeting. He explains that the dissolution, which was unlikely due to Strafford's opposition, became possible because of his absence. This case illustrates an HPE explaining how specific circumstances (in this case, the absence of key individuals) permit surprising historical outcomes.

5.4.1.3 *Revolution of 1688*

In this example, Dray (p. 163) bases himself on an explanation by Trevelyan about how the success of the Revolution of 1688 and 1889 was heavily influenced by chance, particularly on the role of James II, whose policies inadvertently strongly united his opponents. This HPE affirms that the Revolution's success, which appeared improbable given the political disturbance and royal despotism under Charles II, became possible because of the unexpected unification against James II. The "eleventh-hour chance" provided by James II's advent made this constitutional and peaceful transition possible.

5.4.1.4 *The End of Cereals' Protection*

Now referring to Halevy's explanation, Dray (p. 166) shows how it was based on why extending the protection on cereals provided by the 1815 Act was not possible. Dray says that Halevy's HPE shows that given the economic conditions—specifically, the price of corn barely exceeding the threshold—extending protection was impossible. This explanation sets the economic factors that made any extension impossible, thereby addressing the apparent paradox of protection not extended despite potential political will.

5.4.2 Goertz's Historical Examples of Necessary Conditions

The political scientist, sociologist, and philosopher of science Gary Goertz, already cited here for his studies on necessary conditions, offered a list of 150 necessary conditions hypotheses that, when it comes to historical events, get closer to what I call a possibilist explanation. However, as an advocate of causal explanations, Goertz does not use laws to explain. Let us see some historical examples of his "database."

5.4.2.1 *Aldrich Oligarchy*

Brady, Brody, and Epstein (as cited in Goertz, 2003, p. 78) describe, "We wish to show that a necessary condition for the strong, centralized leadership exhibited in the Senate by the Aldrich oligarchy was the existence of a party membership that was homogeneous in its policy preferences."

To explain it in a possibilist way, we should start by saying that there is a generalization that there can be no strong and centralized leadership in a parliament without uniformity and cohesion in the majority party. As this necessary condition was satisfied in the Aldrich oligarchy, its strong regime was possible. If the uniformity among party members facilitated unified leadership because it minimized internal conflicts and divergent interests that could potentially undermine a centralized leadership approach, then another page: party cohesion probably is only a permissive condition, not a necessity for centralized leadership. But the nature of science is the constant falsification of hypothesis and its posterior refinement.

5.4.2.2 *The Falsification of the Indispensability of Railroads*

This example is interesting because it shows how the necessary conditions hypothesis can be falsified. Fogel (as cited in Goertz, 2003, p. 81) writes:

> The axiom of indispensability … the evidence usually cited does not support the traditional interpretation of the imperative role of the railroad in American economic growth. This evidence fails to establish a causal relationship between the railroad and either the regional reorganization of trade, or the change in the structure of output, or the rise in per capita [income], or the various other strategy changes that characterized the American economy of the last century.

We can refurbish this argument as a falsification of a possibilist explanation, saying that for the generalization that says that without the connectivity of railroads, we have no economic development, the evidence about America's development does not show a causal relationship between railroads and its regional reorganization of trade, change in the structure of output, or the rise of per capita income. Therefore, this generalization, not America's economic data, is falsified.

5.4.2.3 *The Beginning of the First World War*

This very well-known episode of our history is an excellent example of an implicit possibilist explanation that took the form of a causal chain of events. Levy (1990, p. 170, as cited in Goertz, 2003, p. 84) defends the explanatory hypothesis that

"German support was a necessary condition for an Austro-Hungarian war against Serbia. The erroneous assumption of British neutrality was a necessary condition for German support of an Austrian invasion of Serbia, and consequently for a continental or world war."

An implicit generalization says that "no country declares war if it has no certainty of its (or its alliance) military superiority." Given that, it is explained why German support was a permissive condition to the Austro-Hungary Empire to declare war on Serbia. The Empire had to be sure that it could count on the German army to face potential retaliation from Russia because of the war on Serbia. In turn, Germany had the certainty of its alliance military superiority against Russia and France, but not also against England. However, it satisfied the permissive condition of them being sure England would not be at the opposite alliance. Under the flawed assumption that Britain would remain neutral, German support was underpinned as a green light for Austria-Hungary to confront Serbia aggressively and, therefore, to the beginning of the First World War. Note that the sufficiency of both decisions was not explained. What was explained was the possibility of deciding that way.

5.4.2.4 *French Revolution*

Goertz (2003, p. 85) presents Mann's hypothesis that "Monarchs granted lawyers corporate privileges, seeking to reduce their social embeddedness. The French monarchy went the farthest, granting patents of nobility carrying material privileges (noblesse de la robe) and rights to corporate assemblies (parliaments). The collapse of their particularistic alliance in the 1780s was a necessary precondition of the French Revolution."

In its attempts to lower the social unrest, the French monarchy awarded lawyers and legal professionals unprecedented corporate privileges. The implicit generalization is that "if the middle-class legal establishment is not embedded with the popular class, no revolution is possible." Still, we can think of this collapse not as a necessary condition but as a permissive one. In this case, there is an upper generalization (the alliance between the middle and popular class) that it has caused, making possible the revolution (but not necessary).

5.4.3 Possibilism in Geography

As we saw in Chapter 1, Paul Vidal de La Blache, a foundational figure in human geography, opposed the deterministic vision in this discipline. To the latter, the natural environment sufficiently determines culture in all its aspects. For what he called, expressly, "possibilism," the natural environment only imposes conditions and restrictions on the possibilities of development but does not sufficiently cause its results.

His explanations are even less structured than those from Dray and are not strictly scientific or possibilist in the way I defined; instead, they are mere suggestions or even hints of it. The four examples I will give here reflect La Blanche's understanding of geography as a field of possibilities where human agency decides how those possibilities are actualized.

5.4.3.1 Olive Culture in Mediterranean

La Blache (1967, p. 215) explains that the "means of sustenance" in the Mediterranean area was severely restricted by the climate conditions, as well as by the soil of southern Europe and North Africa. He describes the cultivation of olives in the Mediterranean as a choice made by the region's people, particularly the Berbers, who face these conditions. To deal with them, inhabitants of the region have developed specific techniques, scales, and economic systems surrounding olive agriculture.

5.4.3.2 Rice Culture in Southeast Asia

La Blache (1967, p. 228) examined, in the same chapter, the rice culture in the Asia section under the influence of the monsoons. Conditioned by the season of torrential rains and terrain with lake-like depressions left by periodic floods of the great rivers of India, few agricultural cultures were able to flourish. However, the choice of rice between the alternatives does not become determined by its natural appearance. An entire culture and innovations required "minute attention" and were labor-intensive. Technological innovations, such as the compartments built to receive the water, "the degree to which the sprouts are immersed in it, the transplanting and replanting of each seedling by hand" (1967, p. 228), all required free choices, creativity, and social collaboration to flourish. The decision to engage in rice cultivation methods varies significantly based on local knowledge, traditions, and social structures.

5.4.3.3 Settlement Patterns in France

La Blache (1908, pp. 30–2) analyzed how geographical features, such as rivers, mountains, and plains, influence the location of cities and towns, due to their strategic and economic advantages. Once founded, its development and possibilities are necessarily conditioned and constrained by the sum of technological level plus geological resources. However, the character, layout, and growth of these settlements are ultimately shaped by the choices and creativity of their populations, which in turn influence the historical, cultural, and political factors.

5.4.3.4 Urban Development in Varied Climates

In *Principles of Human Geography*, La Blache (1967, p. 271) examines how geographical factors, such as climate and locally available materials, condition the architectural characteristics of human settlements. He shows that in hotter climates, buildings have thicker walls and cities narrower streets to mitigate the effects of heat. Despite these environmental constraints, he emphasizes that architectural styles and urban planning ultimately reflect a set of cultural values, technological advancements, and economic conditions possessed by specific people. It is again his view that while the natural environment provides a set of conditions that constrain human actions, the free choices of a community ultimately determine the actual layout and aesthetic qualities of cities.

All these examples are not strict possibilist explanations because they lack the distinction between necessary and permissive conditions and the explicitation of the law or generalization that was instantiated or falsified. They indicate the importance of necessary conditions in historical explanations and how questions of possibility and impossibility explain historical outcomes. It is only expected, as its highly complex causal environments and lack of information forbid even the most naïve determinist to offer sufficient explanations of any event. Historians use to explain historical events not by declaring them inevitable but by demonstrating their possibility within a given historical context.

5.5 Possibilism in Social Sciences

Gary Goertz's work on the necessary conditions is an important element of the development of the possibilist framework. His causally "necessary" approach has significantly enriched the understanding and application of causal analysis in the social sciences. Goertz emphasizes the importance of identifying conditions that must be present for a particular outcome to occur, challenging the traditional emphases on sufficient (or, I would say, "contributory") conditions prevalent in much of social science research.

Goertz (2003) advocates the use of set-theoretic methods, including Qualitative Comparative Analysis (QCA), to identify causal relationships and necessary conditions. He emphasizes the importance of precisely defining concepts before measuring and analyzing them. Nevertheless, more importantly to our object in this book, he has helped develop the concept of "continuous necessary condition" that involves variables that can take on a range of values. Goertz examines how thresholds can be crucial in determining necessity in such contexts. For example, a certain level of economic development might be necessary for a stable democracy. Given this, what will count as a "necessary condition" is the limit that the variable must reach. A continuous necessary condition can be treated as a traditional discrete necessary condition, depending on the precise definition of its thresholds. That does not mean there are no discrete necessary conditions in social sciences where the condition either exists or does not exist. There are. For example, a country must have a written constitution to be considered a constitutional republic.

His explanation approach is causal and makes no explicit use of laws, but he is aware of the negative nature of the necessary conditions hypothesis. Says he (2003, p. 73), "explicit sufficient condition hypotheses are quite rare. That is all the more curious since, in principle, a necessary condition hypothesis can be converted into a sufficient condition via the contrapositive operation, whereby 'Z is necessary for Y' becomes 'not-Z is sufficient for not-Y'." The transformation of such individual conditions in kinds of events universally causally related makes what I call "possibilist law." However, here, these "laws" are scientific generalizations, not metaphysical necessities.

There is no field of knowledge, except psychology, that has more potential benefits from adopting a possibilist point of view. It is in social sciences that we find the most intricate causal environment. However, possibilist knowledge can simplify the work of an empirical scientist in the quest for knowledge by demanding only the establishment of one necessary condition for an event to be possible. Possibilism in social sciences would offer explanations with a simple and precise format since it would be enough, at the limit, to determine a single variable to make predictions or to explain our explanandum: "As x was not present, given the law X, then behavior y was not possible."

Therefore, here, I will use his "database" of necessary conditions hypothesis and show several of them that can be expressed as possibilist laws or generalizations. They will span the fields of sociology, anthropology, and political science.

5.5.1 Possibilism in Political Science

The existence of "laws" or at least generalizations in political science may appear strange to those who see it as a branch of philosophy. However, there are several examples of possibilist laws in their empirical form.

5.5.1.1 Classical Deterrence Theory

The original hypothesis of Deterrence Theory considers "a balance of capabilities as a necessary, though not sufficient, condition for deterrence stability. For peace to reign, warfare must also be excessively costly. In this view, each of the two pillars, bipolarity, and nuclear weapons, must exist before war can be considered untenable" (Zagare, 1996).

According to this theory, a state can prevent adversaries from aggression by demonstrating the capability and willingness to use substantial military force. The credible threat of severe retaliation would maintain stability and prevent conflicts, assuming, evidently, the model that rational actors will avoid actions that lead to guaranteed, unacceptable costs. It was particularly influential during the Cold War, guiding the nuclear "mutually assured destruction strategy."

We can state the possibilist law as "If there is no credible threat of severe retaliation, then stability is not possible."

$$\forall x\ (\neg Tx \rightarrow \neg\Diamond Sx)$$

where

x = single variable that always denotes a token-level instance of international system at some point in history.

Tx = "a credible threat exists in the international system x."

Sx = "deterrence stability is maintained in system x."

Moreover, to a possibilist explanation, we should add that:

In this system, $(Px \wedge Nx) \rightarrow Tx$
where
Px = "a balance of power (bipolarity) exists in the international system x."
Nx = "nuclear weapons are present in system x."
In other words, the balance of power and nuclear power would be, together, permissive conditions for a credible threat to the agents in the system.

5.5.2.2 *Stone's Revolution Theory*

According to Stone (1972, p. 10), "revolution becomes possible when a condition of multiple dysfunctions meets an intransigent elite; just such a conjunction occurred in the decades before the English, the French, and the Russian revolutions. Revolution only becomes probable however, if certain special factors intervene: the 'precipitants' or 'accelerators.'"

Clearly distinguishing between conjointly necessary and contributory conditions, the possibilist law in this theory can be formalized as:

$$\text{Formalization: } \forall x\,(\neg(Dx \wedge Ix) \rightarrow \neg\Diamond Rx)$$

where
Dx = "a condition of multiple dysfunctions occurs in society x."
Ix = "an intransigent elite is present in society x."
Rx = "a revolution happens in society x."
Equally, there is in this little Stone (1972) text a possibilist explanation of the French, English, and Russian revolutions, when it posits that the conjunction of multiple societal dysfunctions and the presence of a rigid, unyielding elite group were present for decades in these countries after their revolutions making a revolution possible. Of course, when we face sociological hypotheses of laws and explanations, it arouses the difficulty of operationally defining these concepts to test them. However, I will not lay down this road here. Neither social sciences nor methodology, in the general sense, are my objects in this book.

5.5.2.3 *Skocpol's Revolution Theory*

Skocpol's (1979) original hypothesis is that "only where there are pressures on states due to wars and international competition and only where these pressures result in a conjuncture of fiscal crisis, abandonment of the state by the dominant classes, and peasant revolts based in strong peasant communities is there the possibility of a social revolution."

$$\text{Possibilist Law: } \forall x\,(\neg(Wx \wedge Fx \wedge Ax \wedge Px) \rightarrow \neg\Diamond Rx)$$

Formalization:
where
Wx = "state x faces pressures due to wars and international competition."

Fx = "state x experiences a fiscal crisis."
Ax = "the dominant classes abandon state x."
Px = "there are peasant revolts based in strong peasant communities in state x."
Rx = "a revolution occurs in state x."
Description:

The possibilist law derived from Skocpol's hypothesis identifies a set of conditions that are conjointly necessary for the emergence of a social revolution. This law asserts that the absence of any of these necessary conditions makes a social revolution impossible.

5.5.2.4 Relative Capabilities Theory

Johnston (1996, p. 256, as cited in Goertz, 2003, p. 83) defended that an improvement in relative capabilities inexorably does not necessarily lead to the initiation of conflict. He said: "Indeed, there are a number of dyad years in Chinese-U.S. and Chinese-Soviet relations where such an improvement occurred and there was no military conflict. Rather, a favorable change in relative capabilities appears to have been a necessary but not sufficient condition."

$$\forall x, y\ (\neg Rx,y \rightarrow \neg\Diamond Wx,y)$$

where
Rx,y = "a favorable change in relative capabilities between nation x and y."
Wx,y = "initiation of military conflict between nation x and y."
Description:

The possibilist law derived from Johnston's hypothesis states that without a favorable change in relative capabilities between two nations, the initiation of military conflict between them does not occur. This does not imply at all that it is a sufficient condition for war. This appears to be an addendum to the theory of the necessary condition of a nation perceiving itself as militarily stronger than others as a necessary condition to declare war. The theory of relative change then regulates those shifts in relations between two nations that, for a period, see each other as a relatively equal military power.

5.5.2 Possibilism in Sociology

Some sociological theories that are possibilist in nature:

5.5.2.1 Social Movements Hypothesis

Oberschall (cited in Goertz, 2003, p. 87) states that: "(1) Discontents and grievances; (2) ideas and beliefs about justice and injustice; (3) the capacity to act collectively; and (4) political opportunity" are necessary causes for the emergence and growth of social movements."

It can be formalized as a Possibilist Law as:

$$\text{Formalization: } \forall x\, (\neg(Gx \wedge Jx \wedge Cx \wedge Ox) \rightarrow \neg Sx)$$

where

Gx = "discontents and Grievances are present in group x."

Jx = "ideas and beliefs about Justice and injustice are prevalent in group x."

Cx = "the capacity to act Collectively is established in group x."

Ox = "political Opportunities are available for group x."

Sx = "a Social movement emerges and grows in group x."

Description:

Oberschall's hypothesis is a complex possibilist law that can be, nevertheless, easily dismembered, as any conjunctive possibilist law, once the absence of any of these factors forbids a social movement's development.

5.5.2.2 *Rivalries Hypothesis*

Maoz and Mor (cited in Goertz, 2003, p. 83) defend the psychological hypothesis, applied to sociology, that "mutual perceptions of capability to win a confrontation are a necessary condition for the development of rivalries."

We can formalize this possibilist law as:

$$\forall x, y\, (\neg Px,y \rightarrow \neg\Diamond Rx,y)$$

where

Px,y = "nations x and y mutually perceive that they have the capability to win a confrontation against each other."

Rx,y = "a rivalry develops between nations x and y."

5.5.2.3 *Functional Anthropology Hypothesis*

The notorious anthropologist Malinowski (1936, p. 133) hypothesized that "the functional view of culture insists therefore upon the principle that in every type of civilization, every custom, material object, idea, and belief fulfills some vital function, has some task to accomplish, and represents an indispensable part within a working whole."

There are two possibilist laws that we can extract from this. The first formalization:

$$\text{Formalization: } \forall x\, (\neg Ox \rightarrow \neg\Diamond Cx)$$

where

Ox = "custom, material object, idea, or belief from a culture x."

Cx = "civilization x functions effectively."

We can add an intermediary possibilist law, that is:

Formalization: $\forall x\, (\neg Fx \rightarrow \neg\Diamond Ox)$

where

Ox = "custom, material object, idea, or belief from a culture x"

Fx = "Vital function of the culture x"

The first law offers a closed interpretation of culture, where every object is indispensable because it is functionally integrated at a given time, in contrast to a more dynamic interpretation of culture, where objects and functions constantly change. The second law expresses the idea that if an anthropologist finds an object (material or immaterial) in some culture, she must look for its vital function in it.

Both formulations capture the essence of the functionalist approach, which defends that cultural elements are not arbitrary but essential in a culture at a given time, fulfilling some communitarian need. Even in this holistic approach, where each part is essential for the system's identity and health, we can extract possibilist laws, but not sufficiency or probabilistic laws. However, we should remember that these possibilist laws are ideal-type representations of functionalist assumptions only.

5.5.2.4 *Equitable Obligations Hypothesis*

Jacobson and Weiss (cited in Goertz, 2003, p. 82) have stated that "for parties to implement and comply with accords, they must feel that the obligations imposed are equitable. The differentiated obligations under the Montreal Protocol for phasing ozone-depleting substances … was a prerequisite for adherence to the accord of a large number of countries."

$$\text{Possibilist Law: } \forall x\,(\neg Ox \rightarrow \neg\Diamond Ax)$$

where

Ox: "A x part to an accord y feels that obligations in it are equitable between parts."

Ax: "A x part adheres and complies with accord y."

With the enunciation of this hypothesis, we see a possibilist explanation for the particular case of why the Montreal Protocol had so many adherents. To the authors, it was because it was perceived as equitable (it was not necessarily objectively equitable), and it was seen like it for the parts because of (permissive condition) "the differentiated obligations … for phasing ozone-depleting substances."

5.5.3 Possibilism in Economics

Economics is one of the most challenging and complex scientific endeavors, and arguably one of the disciplines most vulnerable to ideological influence and self-interested claims disguised as science. That is so because economics deals directly with the form by which society distributes its resources and regulates labor to allocate and produce them. Therefore, economics is not a regular science. Rarely, if ever, will it determine what a society should do, but, if honest, it will only predict the consequences of what may happen given the different allocation options and institutional organization. Moreover, I believe that several contemporary economic "laws" are merely generalizations based on correlational studies and express spurious causal hypotheses.

A society chooses how to organize itself economically based on its values and the power of its interest groups. There is no "scientific way" to do this. The parts of economics that impose real constraints, therefore, are necessary conditions for economic activities. They stem from their connections to mathematics, physics, and the boundaries of human cognition and motivation. That imposes real, non-ideological constraints on productivity and monetary relations, and is where the proper laws and generalizations of it (possibilist ones) reside. Economics can greatly benefit from a possibilistic turn in its approach to thinking and exposition, enabling people to comprehend the actual limits between necessary conditions for producing goods and political interests in their distribution.

Let us examine some possibilistic economic generalizations and use them as case studies for the Malthusian ideology of scarcity to illustrate how possibilism can help economics escape the hands of ideology.

5.5.3.1 Malthusian Scarcity Theory

Thomas Malthus was an influential classical British economist and demographer. He is widely known for his theory on population growth. In his 1798 work, *An Essay on the Principle of Population* (Malthus, 1798/1998), he argues that population growth—unless checked by limits such as disease, war, or moral restraint—would inevitably outpace agricultural production confined to limited lands, leading to widespread poverty and famine. This hypothesis remains, to this day, a guiding principle for birth control policies.

For him, population, if unchecked, always increases exponentially, whereas food production only grows, if so, linearly, until the limit of land availability. This contrast between exponential population growth and limited food supply arises from the physical constraint that land is finite, and agricultural productivity declines as the most fertile areas are utilized. That means that food production can grow in a roughly linear fashion for a time, until it eventually reaches a ceiling (as the best soil and terrain become occupied, which limits the potential for agricultural yields).

We can restate some aspects of his theory in a possibilist way. The most important of them is the prediction that if there is no more new land to crop, then indefinite population growth is impossible. Let us leave this clear formulation:

$$\forall x\,(\neg Lx \rightarrow \neg\Diamond Gx)$$

where

Lx: "A population x has new lands to crop."

Gx: "A population x has no limits to growth."

This theory has a more striking formulation: "If there are no population checks, then it is impossible to maintain a sustainable population level." Its formalization is:

$$\forall x\,(\neg Cx \rightarrow \neg\Diamond Sx)$$

Cx: "A population x has periodic 'checks.'"

Sx: "A population x is at a size that is 'sustainable.'"

Let us add subtitles to these formulas. We should not misinterpret Malthus. He never directly advocated coercion, famine, or war as a means of population control. However, when he discusses "population checks," he objectively considers factors such as disease, famine, war, or increased moral restraints on reproduction. Only that can effectively limit or reduce population growth. From these, two can be natural, but two are inherently unnatural. Famine and disease will necessarily follow from sub-nutrition, eventually caused by the incapacity of producing more, but war and moral restraints on reproduction must be artificial. Following this line of thought, artificially spreading disease, causing famine, raging a war, or inventing moral restraints for new human beings to live by becomes morally justified. But are the premises of his theory true?

Is it true that natural resources are limited? Or at least land? Moreover, is it true that if natural resources are limited, then our production capacity will be limited? The answer is no; both premises are false. Natural resources would not be limited if we were to explore the entire universe. Similarly, our production capacity would not be limited if our technological advancements were not limited.

As fantastic as it may seem, unlimited technological advancement would allow us to desalinate water and convert it into agricultural land in the Sahara and Arabian deserts. That allows us to produce hydroponic food in multistory buildings today. Even more, new technology would enable us to reach the Moon, Mars, and eventually beyond, addressing our need for more land, space, and natural resources. Even in the scarcity of new lands and mineral resources, unlimited technological advancements would give us the power to alter matter (today, we already create artificial diamonds) and access unlimited energy sources. In both cases, what imposes the limit to our growth is a limitation in our scientific and technological development, not natural resources. The positive formulation for this is that if our science and technology continue to increase indefinitely, then our potential growth also rises indefinitely.

In a world where change was scarce and slow, and technological advancements were also limited, Malthus could not foresee the impact of technological advancements on improving productivity and better use of natural resources. The developments in the centuries following the formulation of his theory dramatically increased the demographic potential, with advancements in agricultural engineering, biology, medicine, and industrial machinery that enabled unprecedented population growth without the catastrophe he had predicted. These historical facts imposed an undeniable falsification on his economic theory. However, scarcity ideology has survived. It influences most of what we call today "environmentalism," "ambientalism," or "sustainability." The loss of biodiversity, water pollution, and climate change is real. They are, among others, critical problems raised by the misuse of technological advancements. However, their solution should not be to stop population growth (especially in developing countries) through violence, economic coercion, and despair. Instead, the solution to these challenges is more technology, used with responsibility to mitigate or eliminate the consequences of its misuse.

5.5.3.2 Economic Sustainability Theory

It is the hypothesis that the physical availability of resources constrains economic activities. The extraction rate of non-renewable resources, such as fossil fuels or certain minerals, cannot exceed the natural replenishment rate (which is zero for non-renewables). A classical formulation of this hypothesis was defended by Nicholas Georgescu-Roegen (1971). More recently, it has been refurbished by Herman Daly (1996). We can express it as "If the rate of resource extraction surpasses the sustainable yield, then the resource base cannot support long-term economic activities." Reframing it, we have the possibilist law:

$$\forall x\ (\neg Sx \rightarrow \neg \Diamond Lx)$$

where

Sx: "x is an economically sustainable resource."

Lx: "x supports long-term economic activities."

In other words, if some resource x is "renewable," humanity can use it indefinitely to produce goods. If it is not, the long-term activity relying on it is already condemned.

This principle is appealing because it seems to be grounded in physical reality. Once a non-renewable resource is finished, by definition, it can no longer support ongoing economic activity. However, this is only the Malthusian principle repaginated. Depending on its nature, scientific and technological innovation can make a non-renewable resource renewable or sustain the same economic activity using alternative energy or material resources. The non-petrol-based plastics and the synthetic production of diamonds are only two examples of many. Therefore, this economic law overstates the criticality of physical constraints by underestimating, again, the transformative power of technological innovation. The reason is that, strictly speaking, a resource is only considered non-renewable given a certain stage of scientific knowledge and technology.

5.5.3.3 Energy Efficiency Theory

"Energy cannot be created or destroyed; it can only be transformed," says the first law of thermodynamics. One of its consequences, therefore, is that the economic output of an energy source is directly impacted by the way it is used. The more inefficiently it is used, the less amount of work we can get from the same energy source. One example is the energy waste resulting from heat loss. It is again Georgescu-Roegen (1971), one of the authors, who defended this simple idea. He formulated the "entropy law of economics," emphasizing irreversible energy degradation and placing strict biophysical limits on economic processes.

Interpreted as a possibilist law, in simple terms, it states that if there is no improvement in energy efficiency, it is not possible to increase the economic output derived from the same amount of energy input.

Formally: $\forall x\ (\neg Ex \rightarrow \neg \Diamond Ix)$

where

Ex: "x is an energy source which has improved efficiency."

Ix: "x increases the output from the same amount of the source."

This law presents a physical limit: how the efficiency of energy conversion directly influences the productivity of an economy. However, while no amount of policy can surpass the intrinsic limits set by thermodynamics, scientific and technological advancements have the potential to do so. Technology can always change the game. It can always improve the efficiency of resource use or find alternatives to its use. For example, renewable energy technologies are helping to reduce our use of fossil fuels. We now generate energy through various sources, including sea waves, solar panels, and advanced wind turbines. In the future, it is not absurd to consider that some atoms could solve our energy problems for an entire year.

5.5.3.4 *Potential Populational Density*

A better understanding of the relationship between technology and sustainability can do justice to the basic intuition and motivation not to destroy our planet or to preserve it by killing people. Our potential to sustain a larger population on this or any planet will be limited not only by our natural resources but also by the level of scientific and technological knowledge we currently possess. Therefore, orienting our policies to put technological advancement before population growth would be prudent. However, the second is only a political, not scientific, value.

We can express this as a possibilist law:

$$\forall x\ (\neg(Tx \lor Nx) \rightarrow \neg Gx)$$

where

Tx: "population x had a Technological advancement."

Nx: "population x had improved access to Natural resources."

Gx: "population x had a Growth in potential populational density."

Over the last few years, innovations such as genetically modified organisms, desert agricultural techniques, and robotics manufacturing processes have contributed to a significantly increased capacity to sustain a growing global population. Moreover, they did so with less impact per capita on the planet's natural resources. That means we should always pursue technological advancements that enhance our potential to sustain population growth on this planet or others. It is possible. If it is desirable, it is a political and moral decision, rather than a scientific one.

5.5.3.5 *Coal as Necessary Condition to Industry*

A last example of this discussion is from Goertz's 150 necessary conditions hypothesis. He collected (2003, p. 76) Bairoch's hypothesis that "England was one of the three or four European regions where coal had already been mined for a long time. Therefore, very likely a factor almost certainly necessary for industrialization, but without a doubt not sufficient."

Alleged possibilist Law: $\forall x\ (\neg Cx \rightarrow \neg \Diamond Ix)$
where

Cx: "Region x has a vast offer of coal."

Ix: "Region x achieves industrialization."

There is a mistake in this law. It was not really a law, because it is historically dependent on holding. The necessary condition satisfied then was not coal; it was a sufficient energy source. Rethinking those allegations in light of our previous discussion, we can see clearly in the possibilist explanation that coal was a permissive condition for industrialization in England because it satisfied the necessary condition of energy source to move the new industrial machinery. Given the technological knowledge of the time, coal was the only viable option for such a large amount of energy, so it served as a necessary condition at that time. However, this is no longer true nowadays.

5.6 Final Remarks

This chapter thoroughly explored possibilist laws across a diverse range of scientific disciplines, illustrating their theoretical foundations and practical applications. This chapter corroborates that possibilist laws are not merely theoretical constructs but have significant empirical relevance, as illustrated by several examples from diverse fields, including physics, biology, history, geography, sociology, and economics.

I have analyzed scientific theories and fields, such as Ohm's Law and Quantum Physics, through a possibilist lens, underscoring the importance of recognizing and understanding the necessary conditions underlying various phenomena as a new and fruitful way of thinking about old problems.

Moreover, the use of possibilism as an instrument to analyze discredited theories, such as Malthusian economics, has shown that it can help distinguish between necessary scientific conditions for economic activity and moral or political reasons for organizing our society economically in a specific way.

All these possibilist laws presented in this chapter provide robust evidence that possibilist knowledge has always been a reality and, if actively and consciously pursued, could help us understand and predict our world. Nevertheless, in the next chapter, I show its most compelling case study: psychology.

6

Psychology and Possibilism

This chapter is a comprehensive case study. It begins by describing psychology's remaining foundational obstacles to becoming a mature, modern science, setting the framework for evaluating the potential impact of possibilism on it. Thus, the hypothesis that there are actual laws in psychology will be justified. They are possibilist laws (PLs) and comparative ceteris paribus laws. Plenty of examples of them will be given. Next, the characteristics of sufficient deductive-nomological and mechanistic psychological explanations will be explained in the context of psychology. Finally, it presents possible explanations of the discipline. The final section shows how possibilism addresses and mitigates several obstacles to scientific psychology, offering a new path toward its unification.

6.1 Obstacles to Psychology as Modern Science

Psychology is closer to becoming a proper modern science than economics, but it does not mean it will be easy. It does not suffer as much as economics from political propaganda and monetarily motivated discourses. However, it deals more closely with the advocates of two ideologically motivated groups and their conceptions of human beings: religious and materialists.

Sociology and economics are, by definition, more complex and challenging than psychology because they study the results of interactions between uncountable "psyches." In other words, their object is the outcome of the interplay of an enormous quantity of psychological objects. Moreover, these objects happen to be free agents (whether with a negative or positive sense of freedom). That is reason enough to affirm that the future of social sciences is being decided in psychology.

Therefore, let us examine the obstacles to psychology in its quest to become a modern science. Based on a previous taxonomy (Castañon, 2009), these obstacles are categorized into ontological, epistemological, methodological, and political domains. I will dedicate a considerable number of lines only to the problems that are still blocking the discipline and only briefly mention those that have been surpassed throughout the history of the discipline.[1] Setting this scenery will better show the potential impact of possibilism on it and in neuroscience.

6.1.1 Ontological Obstacles to Psychology

By "ontological," I mean the questions about the object of the discipline, its definition, and nature. They are the most fundamental and challenging allegations to overcome, as the mind's subjective, non-observable, creative, and probably free nature. Most of the formulations of these problems are derived, in one way or another, from Immanuel Kant in the *Critique of Pure Reason* (1781/2001) or the preface to the *Metaphysical Foundations of Natural Science* (1786/2004).

6.1.1.1 Unquantifiability

Kant posited direct vetoes to the possibility of psychology becoming an empirical science. The first was about the supposed impossibility of quantifying psychological phenomena. Psychology as an empirical science would neither proceed a priori nor be able to quantify its data and employ mathematical calculation in the precise description of reality and the laws that govern it. This veto would be definitive: psychic phenomena occur only in time, not in space. Therefore, they would not be capable of quantification. Kant (1786/2004) says on this issue that "mathematics is not applicable to the phenomena of inner sense and their laws; the only option one would have would be to take the law of continuity in the flux of inner changes into account" (p. 7).

Behaviorism was the first to offer a consistent solution to this problem, making behavior, which occurs in space, the object of psychological study. The problem with this solution is as well-known as the solution itself. Despite being the only source of objective data, this object is not the primary object of interest for most psychologists. In the mid-twentieth century, cognitivism addressed this problem, enabling the scientific method to investigate genuine psychological phenomena through two innovations.

Often not explicit or conscious in the field's research strategy, the first innovation was the replacement of mathematical descriptions of laws in psychology with merely logical descriptions. The problem of scientific descriptive rigor is thus faced with a language that is as precise as mathematics but not quantified. Baars (1986) refers to this language as "non-quantitative mathematics," encompassing symbolic logic, Boolean algebra, topology, and recursive function theory. Computerized simulation proved particularly useful in this regard, as it provided an adequate and sufficiently explicit language for constructing cognitive models and hypotheses precise enough to be falsifiable. Possibilism can also add one more brick to this house, and we will return to this.

The second was methodological. By starting to use chronometric data to test hypotheses about cognitive processes, cognitivism reversed Kant's fragility condition with the simple observation that if cognitive functions occur over time, this time can then be measured through the duration of task execution.

6.1.1.2 Indivisibility

According to Kant (1787/2001), the psychic phenomenon is indivisible; it is impossible to investigate it through analysis and synthesis, as psychic events cannot be considered separately as elements. This veto was developed by several philosophers, including Franz Brentano and Edmund Husserl, and influenced the Gestalt approach in psychology.

Brentano (1874/1995) argues that the intentionality of consciousness is the distinctive characteristic of the uniqueness of mental life and advocates for a psychology of the mental acts. Husserl (1911/1952) argues that psychology, as a rigorous science, can only be a form of philosophy, specifically, a phenomenology. Psychology should be a science of the essences of intentional mental acts.

The indivisibility obstacle remains (Dreyfus, 1993) because it is ontological in nature. However, the phenomenological tradition has overestimated its importance. Psychological phenomena are artificially analyzed into parts—just as biological and physical systems are—though such divisions are less problematic in those domains. The organism must also be seen as a holistic, indivisible whole, just like the universe. The consideration of organic and astrophysical problems in parts, although artificial, is a condition for their study. Psychological life is unitary, but functions and representations can be scientifically investigated in isolation. Having assumed scientific knowledge as approximate, cognitivism abandons the illusion of absolute knowledge from the beginning, which means that this veto becomes just a constraint, not a prohibition.

6.1.1.3 Meaning as Psychology's Object

For humanistic psychology, psychology's object should be the meaning of conscious experience. It intends to use experimental methods to investigate this, together with idiographic ones. The existential psychiatrist Viktor Frankl (2006) defended this position, arguing that the investigation of the human being in terms of cause and effect is distorted and incomplete, as it lacks consideration of the meaning that one gives to one's own life.

From a humanistic perspective, the human being is not the result of a series of things, but rather the beginning of a series of things. The experimental approach refers to the human being as a result, as past; it does not refer to it as present, challenged by questions of meaning. Therefore, the goal of psychology is to provide a comprehensive description of what it means to be alive as a human (De Carvalho, 1990). The plain problem with this is the merely qualitative nature of this type of object, unsuitable to the scientific method. That will generate an epistemological obstacle evaluated on 6.1.2.3.

This radical postulation overlooks the fact that information processing functions are fundamental and integral psychological phenomena. As Fodor (1991b) postulated with his principle of "methodological solipsism," scientific research is restricted to representations and their transformation functions, and we can predict the behavioral result of an informational input without resorting to its meaning. We may have a dead end here: if meaning is a psychological object, it cannot be scientifically investigated; if not, psychology has no meaning (literally). May possibilism help with this constraint of psychology as a discipline? I think so. We will come back to this.

6.1.1.4 Human Being as Autonomous

For much of Western thought, humans are endowed with some self-determination; we are sources of activity in the universe, not merely reacting objects. The classic formulation of this problem was the Locke-Leibniz controversy. After Locke (1689/1978) published his "*Essays Concerning Human Understanding*," Leibniz (1704/1974) wrote his "*New Essays*," setting the terms of the modern debate on the origins of knowledge and human freedom. For Locke and empiricism, human beings are passive, a blank sheet of paper to be written on by experience, a receptacle of sensory impressions that will shape their intellect and govern their actions. Leibniz, on the other hand, describes human beings as being influenced by these perceptions, yet free, active, and purposefully oriented toward goals—a focus of activity in the universe, not a passive object driven by physical forces.

Allport (1955/2000) argues that these two opposing models have generated two traditions in psychology: the Lockean and the Leibnizian. According to Rychlak (1975), these traditions are Lockean and Kantian. To those "Lockean" schools of psychology that assume determinism, this problem is addressed by behaviorism, computationalist cognitivism, and biological psychology. However, for the common-sense and "Kantian" schools—including gestalt, phenomenological, and humanistic psychology, methodological cognitivism, and Piagetian constructivism—there is some level of autonomy, and the problem remains.

The consequence of a creative and free consciousness is that human beings can be radically unpredictable. And if people vary widely in their behavior and psychological processes, it becomes harder, or even impossible, to establish universal principles that apply to everyone. The last radical conclusion is defended by anti-Kantian and postmodern approaches such as Social Constructionism (Gergen, 1973).

This issue is the central cause of psychology's fragmentation, the greatest obstacle to the unification of this science, and, to date, it remains without a philosophical solution that at least allows for a minimum methodological compromise. We will explore how possibilist knowledge can help solve this problem.

6.1.1.5 No Proper Object

The final claim of the physicalist tradition, eliminative materialism (Churchland, 1981), and social constructionism (Gergen, 1973) is that psychology has no proper object. To be a distinct discipline, psychology must have an object that resists the reduction to the object of biology or physics. Ontological reduction in psychology can be defined as the process of reducing all psychological phenomena to a single non-psychological causal principle, which remains unalterable throughout the phenomena's transformations. The reduction in science (Hempel, 1966) consists of considering that specific orders of phenomena are subject to better-established and more precise laws of another order of phenomena that is ontologically more fundamental, such as, for example, that organic phenomena are derivable from physical laws (mechanism versus vitalism). Reductionism in psychology generally holds that all psychological phenomena are fundamentally biological. But not always. It can be behaviorist (there is only behavior) or sociological (there is no subject, only language).

I will not enter more into this issue because this obstacle is no longer widely considered. Since the advent of the computational metaphor (Neisser, 1967), originating from Putnam's (1961) functionalist theses, psychology has offered an answer to this problem that eliminates the specter of a lack of proper object that had threatened it since its foundation. For functionalism, reductionism is often inapplicable to science because its objects result from the emergent properties of complex organizations of more basic constituent elements. Thus, psychological functions are emergent properties of the brain, just as software is an emergent property of specific hardware organizations. Understanding a given pattern of information processing does not require understanding how it is being physically processed. Thus, at least regarding cognitive functions, even physicalists must concede that psychology would have an object that is irreducible to objects from other disciplines.

6.1.2 Epistemological Obstacles to Psychology

By "epistemological" here, I mean the questions related to the process of acquiring knowledge, its origin, and justification. However, they are intrinsically linked to limitations due to the nature of the object of the discipline. Some of them are also derived from Kant's critics.

6.1.2.1 The Subject of Knowing Cannot Simultaneously Be Its Object

This Kantian veto against empirical psychology is also presented in the "paralogisms" of the *Critique* (1781/2001) and the preface to the *Principles* (1786/2004). Internal observation "by itself already changes and displaces the state of the observed object," which condemns psychology to be a "natural description of the soul" (2004, p. 07) but never an experimental discipline.

Auguste Comte assumed this veto. The thinking subject cannot be at the same time the object of the experiment or even of the mere observation that he performs, since he would be aware of the experimental and control conditions, in addition to the internal observation interfering in the natural outcome of the psychic process itself. According to him, every active being can only be studied statically, in its conditions of possibility for action, and dynamically, in action. In the first form, we would study the organic conditions on which psychic phenomena depend: physiology. In the second, dynamics, we study "the effective march of the human spirit in action" (Comte, 1973, p. 19): Sociology.

Penna (1991) believes that Comte's veto on psychology results from its definition of a science of the soul based on introspection, which was inadmissible for positivism. However, Comte himself would be convinced that something like "positive psychology" (the works of Fechner and Weber) had just been founded in his time, whose methods he approved.

All solutions to this problem come from behaviorism. Making behavior the object of the discipline placed the object to be investigated outside the investigator. Even the cognitivist approach to this problem is due to a dissident of behaviorism, Edward Tolman. His concept of intervening variables enabled us to infer cognitive processes as functions that transform environmental information entering the organism into behavioral information exiting it. Another noteworthy fact is the new approach to introspection in cognitivism, where it loses its status as a hypothesis test and begins to have a merely descriptive and exploratory function in the form of verbal protocols.

6.1.2.2 Knowledge and Interaction Alter Psychology Object

The problem of the alteration of psychology's object through interaction and acquired knowledge was raised by sources such as Kant (1786/2004), cognitivism (Neisser, 1975), social constructionism (Gergen, 1973), and humanistic psychology. It defies the assumption of the object's regularity.

Gergen (1973) argues that knowledge has social consequences: the dissemination of psychological theories modifies the patterns of behavior on which the theories were built, for some reasons. One of them would be the high credibility that science enjoys in our society, causing many people to guide their conduct according to psychological theories. Conversely, some people who see their beliefs and values attacked by specific theories may exercise the will to affirm their values by challenging scientific predictions. Moreover, disseminating scientific theories changes behavior patterns by promoting individuals' emancipation from factors that conditioned their behavior and were unknown to them. Moreover, all this is only part of Gergen's story. As a relativist, he views psychological phenomena as completely context-dependent, determined by cultural and social factors that preclude generalizations.

In Cognition and Reality, the cognitivist Ulric Neisser (1975) notes that we must consider the apparent paradox that the increase in knowledge leads to an increase in the unpredictability of behavior: the more we know about the environment and our cognitive processes, the less predictable our behavior becomes. He argues that we cannot predict a person's behavior in a natural environment unless we know as much about it as the person knows. To Neisser, a psychologist is not the best person to predict the movements of a chess master in a game because he does not know as much as the master; another master is better suited to make predictions. If the behavior is a function of information available in the environment, the more someone understands her environment, the more information she will gather from it, and therefore, the more difficult it will be to predict her behavior: "Truth does make us free" (Neisser, 1975, p. 185).

Despite all this, we should not interpret this epistemic obstacle as an ontological veto for scientific investigation, but rather as an epistemic complication. If a pattern governs the change during interaction, this is not evidence for indeterminism. On the contrary, we would have evidence of some laws regulating the object's interaction with others. Cognitivism also bequeathed to psychology a mathematical model for dealing with these processes, the feedback model. Furthermore, methodological procedures attenuate this problem, which also affects medicine, by hiding information about the groups and hypotheses being tested from both subjects and some researchers, such as the double-blind procedure, the placebo group, and even deception in research. Today, the allegations surrounding this supposed obstacle have been largely abandoned. However, possibilist knowledge has the potential to bury it in its most radical formulation. We are about to see it.

6.1.2.3 *Psychology Offers Another Kind of Knowledge*

If the object of psychology (consciousness) is ontologically different from the object of the natural sciences (physical objects in space), then it requires its own method of investigation. This is the thesis of Husserl (1911/1952), who proposes phenomenology as its proper method. Husserl (1936/2002) states that modern science objectifies consciousness, creating an artificial object that has no relation to the real one. Intentionality distinguishes consciousness from phenomena of physical order. Consciousness is always consciousness of something and never an object; it transcends itself by nature.

It is also the thesis of Dilthey (1924/1945), who proposes classifying psychology as a science of the spirit with an interpretative and comprehensive method, rather than an explanatory one. Psychological phenomena are unique, unrepeatable, and need, instead of discovering the causes or laws that govern them, to be understood in their individuality, with the reasons for them.

Criticisms such as those of Husserl and Dilthey have no definitive solution. As Searle (1992) argues, the approach to consciousness and the unconscious as third-person phenomena does not exhaust them. Therefore, we cannot restrict their study to the scientific method. But we cannot abandon it either. Our challenge is to apply the most effective method for acquiring knowledge about nature to psychology, since the human psyche is a part of nature. I am about to offer a compromise solution to this problem. This solution, however, depends on recognizing the limits of scientific psychology.

6.1.2.4 *Mental Contents Are Not Directly Observable*

The oldest limitation of psychology's object is that it is not observable through our senses. And for old naive empiricism, as the senses were the source of all knowledge about the world, the mental processes cannot be objects of scientific investigation. Nevertheless, even if we do not consider the senses as the sole source of knowledge, their role remains as a source of observational propositions that will judge our hypotheses about the world. How can we judge hypotheses about psychological phenomena if we do not have public, objective observations of mental processes?

Critical rationalism (Popper, 1959/2002) posits that the scientific method is hypothetical-deductive, not inductive. This approach overturned the myths of the scientific need for direct observation of phenomena and the notions of induction and neutral observation. Today, it is clear that a scientist creates a theory, deduces an observable consequence from it, and performs an experiment to test the prediction. Therefore, this obstacle is gone. Psychology's object of study is as unobservable as that of quantum physics. It hypothesizes cognitive processes and structures, predicts the behavior that will derive from them given some informational input, and tests its prediction by measuring informational outputs through controlled experiments.

6.1.3 Methodological Obstacles to Psychology

There is no clear frontier between epistemological and methodological issues. Epistemic choices underpin methodological ones. Epistemic issues concern the foundations of knowledge, while methodological issues are more specific, practice-focused, and concern the execution of research and inquiry. The methodological obstacles to psychology arise from the tentative application of experimental methods to this discipline, primarily concerned with the practices and procedures used by those investigating psychological issues.

6.1.3.1 Ethical Limitations

It is difficult to overcome the limitations imposed by ethical imperatives on using human beings as research objects. Often, the appropriate scientific forms of measurement and control cannot be imposed on human beings. Thus, to study the effect of certain events (generally harmful), we must wait for them to occur by chance or gather evidence from isolated cases. This is a powerful obstacle to advancing psychology.

There is yet another difficulty in controlling psychology's object: the experimenter's effects on the object are significant. It may result from a conscious or unconscious attempt by the experimenter to bias the experiment's outcome in favor of confirming their personal beliefs or theories.

Despite limiting the speed of psychology's development, ethical care in psychological research is no different from that required in medical research. That does not make modern psychology impossible but difficult. The experimental designs become more demanding of scientists' creativity and patience.

6.1.3.2 Measurement Limitations

Any philosophy of psychology must define a satisfactory approach to measuring data from psychological research. Psychological constructs (like intelligence, motivation, or emotion) are abstract and often lack even philosophically precise definitions, leading to challenges in their measurement and operationalization. The validity (the property of accurately measuring the intended object) of operational definitions for these constructs is often subject to harsh criticism because these definitions are usually too artificial. Not to mention that there are many operationalizations of psychological phenomena, leading to inconsistencies and difficulties in comparing results across studies. Moreover, ensuring that psychological measurements are reliable (i.e., consistent across time and different contexts) seems more challenging for psychology when dealing with bias from both subjects and researchers. For humanistic psychologists, psychology has even failed to apply a mathematical treatment to research and problems because its object is not quantifiable.

We should question the radicalism of dispensing data measurement, as the basis of objectivity and precision has given the other sciences their incredible progress. Measurement is not the objective of the experimental psychologist; it is a tool for investigating reality. Moreover, if it is more artificial in psychology than in any other biological science, psychologists have developed it creatively since the discipline's beginning.

In any case, since the disidentification between knowledge and certainty and the dissemination of the hypothetical-deductive method, this problem has been alleviated. Thus, psychologists could begin creating measurement instruments for the effects of cognition that their theories and hypotheses, under test, deemed valid. Today, an immense variety of these effects are measured, some of them directly linked to the psychological processes they represent, such as the duration of task execution, reaction time, eye movement, various types of physiological reactions, the brain's behavior through its blood flow, in addition to the traditional response acuity data from self-reports.

6.1.3.3 Overwhelming Quantity of Variables

Psychology deals with an immense number of variables (biological, cognitive, motivational, emotional, social, environmental, and informational) that interfere with the determination of its phenomena. These variables interact in complex ways, making it challenging to isolate the effect of any of them.

We evaluate the complexity of a science by the number of variables intervening in its object of study. In psychology, they are overwhelming, increasing the chance of misuse or misinterpretation of statistical data. One common error is due to confounding variables that may influence both independent and dependent variables, leading to spurious associations. Another comes from interrelated variables, when variables are correlated (such as stress and anxiety), making it difficult to determine the effect of each one.

Even from a reductionist perspective, we conclude that it is necessary to know a lot about chemistry to understand the action of neurotransmitters, which is necessary to understand neural functioning, which is necessary for explaining the highly complex brain functioning, which, ultimately, would be the cause of human behavior. Furthermore, when considering human beings' social relationships, we face virtually innumerable informational variables influencing their behavior.

If we also adopt an approach in which humans are free and proactive, we face an immeasurable and unpredictable complexity scale. In this chapter, I have yet to demonstrate how possibilism, complemented by the causal interpretation of ceteris paribus laws, primarily addresses this issue.

6.1.3.4 Time Span

A considerable difficulty of psychology is its reliance on longitudinal studies for understanding developmental and long-term psychological processes. Longitudinal studies enable researchers to observe individuals over extended periods.

Practical issues frequently impact these studies. The obvious is its extended duration, which often spans many years or even decades, allowing for the tracking of development and changes in psychological processes. It requires a high budget (guaranteed over the study) and prolonged commitment from researchers and participants. Moreover, it is a feature that makes progress extremely slow, emerging questions too challenging to address within the same study, and replications too expensive to be worthwhile.

It is pretty common to lose researchers and participants over time. Retaining personnel and participants can be problematic for various reasons, including loss of interest, relocation, illness, or death. Another unexpected problem is significant changes in science and technology over the years, which have the potential to introduce new intervening variables and changes in measurement tools. These limitations are not impeditive, just challenging, and are cited here to compose the framework of psychology's difficulties. Possibilism has nothing to help it.

6.1.4 Political Obstacles to Psychology

It may seem odd to some to consider "political" obstacles to modern psychology in this context. Of course, they are always entangled with theoretical problems. However, predominantly political obstacles to psychology's development arise from government policies, legislation, socio-political climates, or actions by interest groups. They can impact funding, diffusion, collaboration, regulations, and public perception, affecting the progress of scientific research and its application. Here, I will address only one of them: the fragmentation of psychology.

6.1.4.1 Fragmentation

Psychology has been the history of various schools and theoretical perspectives, each with its own methods and objects. From its beginnings with structuralism, passing through functionalism, associationism, psychoanalysis, gestaltism, behaviorism, humanism, and cognitivism, psychology has been a battlefield of anthropological beliefs. More than just a problem of "diversity" or "complexity," each school has its own assumptions and conceptual structure, often accompanied by allegations of "revolutions" or "paradigm shifts" (after Kuhn, 1962). The proliferation of psychological approaches leads to different and incommunicable methodologies, language, and results, making it impossible to build cumulative knowledge. This Babel restricts communication within psychology and integration with the conjunction of the sciences, severely impairing the intersections of psychology with biology and neuroscience.

Adding complexity to this enormous problem, we have another effect of the object's subjectivity: the apparent difficulty in adopting scientific attitudes free from prejudice about an object, which, ultimately, is ourselves. The subject matter of psychology is more subject to the projection of our ideas than that of other sciences. Popper (1979) argues that this is a characteristic of the empiricist tradition since prejudices are constituent and inseparable parts of scientific activity. Furthermore, the object of natural sciences would be no more deprived of our projections than that of psychology. Nevertheless, by dealing with a vision of ourselves, psychology has revealed itself to be the discipline with the most extreme and irrational reactions to experimental evidence, a rich source of examples of what Bachelard (1974) called an "epistemological obstacle."

6.1.5 Psychology's Present Status as Modern Science

We have seen a formidable set of obstacles to the full constitution of scientific psychology as a modern science. Most do not preclude psychology but only add difficulties to it. Others were significantly attenuated or methodologically overcome and are now outdated.

Among the significantly attenuated obstacles, the subject being at the same time its object (6.1.2.1). We have long since abandoned introspection, turning our source of data collection into overt behaviors. The psychic phenomenon is indivisible (6.1.1.2), as any existing phenomenon. This ontological limitation is today methodologically overcome through cognitive decomposition and model-based simulations. Analyzing unlimited phenomena is a necessary methodological resource for considering them within our limited mental capacity. As our knowledge is approximate, minor distortions resulting from these analyses can be tolerated. The thesis that psychology does not have an object (6.1.1.5) is decidedly mistaken even from a physicalist point of view since it ignores the distinct order between the physical (the brain) and the functions it performs (the cognitive processes). Likewise, there is no more sense in requiring direct access to the object through the senses (6.1.2.4). Neither physics, astrophysics, nor biology has today directly observable objects of investigation (e.g., bosons, black holes, and DNA).

Methodological obstacles do not preclude scientific psychology. Ethical limitations (6.1.3.1), measurement difficulties (6.1.3.2), and time span (6.1.3.4) are operational issues that only add costs, duration, mistakes, and difficulties to psychological research. Similarly, political obstacles are not transcendental but rather objective conditions of our society.

Nevertheless, the remaining ontological and epistemic obstacles are sufficient to consider the possibility of a scientific psychology sub judice. However, possibilism can offer solutions or attenuations to unquantifiability (6.1.1.1), meaning as an object (6.1.1.3), the autonomy of human being (6.1.1.4), alterations in interactions (6.1.2.2), and the idiographic dimension of psychology (6.1.2.3). Moreover, there are two tremendous obstacles to psychology, one methodological and the other political, that possibilism can also attenuate. They are the quantity of variables in psychological research (6.1.3.3) and the discipline fragmentation (6.1.4.1). This fragmentation, often misunderstood as disciplinary pluralism, reflects not a division of subfields, but a competition of incompatible paradigms. Possibilism may provide a conceptual framework for reconciling or coordinating them. These reflexes will be addressed next, beginning with a detailed articulation of how PLs can support psychological explanation without requiring quantification or determinist causation.

6.2 Psychological Laws and Possibilism

Even today, the term "psychological law" can evoke manifestations of skepticism or even sarcasm. Here I will offer a summary of the debate about laws in the discipline, surveying five approaches: (i) law skepticism (Davidson, Rosenberg); (ii) the defense of ceteris-paribus laws (Fodor, Marr, Pylyshyn); (iii) statistical/propensity "laws" (Meehl); (iv) mechanistic "effect laws" (Cummins); and (v) interventionist invariance (Woodward).

Influential philosophers like Donald Davidson (1980) defended the thesis that it is impossible to have strict psychological or psychophysical laws, as there would be no way of specifying the conditions for a universal proposition to be fulfilled without exceptions. To him, intentional mental states are inherently normative and interpretative, which makes them unsuitable for the kind of precise, quantitative laws characteristic of science.

Rosenberg (1996) summarizes two major arguments against psychological laws in *"Laws, Damn Laws, and Ceteris Paribus Clauses."* First, natural "selection for function is blind to differences in structure" (p. 192), so any functional kind (belief, desire, pain, etc.) is realized by "structurally heterogeneous" physical systems. In other words, there are multiple realizations across multiple structures of the same psychological function, therefore, no psychophysical law is possible. This idea is repeated elsewhere (Rosenberg, 2012, p. 189):

> there are so many different ways that the same higher-level fact can be realized by, instanced by, composed of lower-level facts, that the higher one can't be reduced to the lower-level one. It's multiply realized by a vast "disjunction" of different ways people's behaviors can make it up.

Second, according to Rosenberg (1996), there are no strict laws at the fundamental level that would be necessary to identify independent interfering factors, allowing the establishment of ceteris paribus clauses. Without these strict laws, there is no way "to provide the truth conditions for the existence of independent interferers" (p. 202); therefore, there are no ceteris paribus laws either. To Rosenberg, the distinctive features of intentionality and functional character of psychological explanation ensure that no genuine psychological laws, strict or ceteris paribus, can exist. The regularities we observe are local, fragile, and non-refinable; they cannot satisfy the predictive and explanatory standards of the physical sciences.

Generally, many scientists and academics, although believers in the powers of modern science to describe and predict the world, do not categorize psychological phenomena in this manner because human behavior and mental processes are too complex, and people and their historical and cultural contexts are too diverse to be subsumed under simple laws.

Outside science, the denouncers of psychological laws are also numerous and diverse. Some libertarians argue that free will and creativity render individual differences too significant and behavior too unpredictable to be accommodated by laws. Others, postmodernists, argue that cultural and linguistic backgrounds construct reality and the so-called "subjects," making psychological laws pure chimeras. Everybody seems to agree only to deny the possibility of such a thing as a psychological law.

Nevertheless, the show must go on in psychology. It makes no sense to think of a mature science that could not offer any kind of law. The primary business of modern science is prediction. The mere correlation between kinds of conditions does not allow anyone to predict anything. To do that, suppose that the relation between these kinds of events found through manipulation is a law that will apply to all relations between those two variables in the same context.

That is why Jerry Fodor's (1974, 1975, 1991a) ideas on the topic were so successful in psychology. His functionalist perspective gives an entirely different approach to the thesis of multiple realizability. Take the mental kind of "pain," for example. It can be realized by several physical structures (a human brain, a Martian's silicon-based brain, or even an android). These structures could all, in principle, be in a state of "pain."

The problem of multiple realizability, posed by Rosenberg (1996), which denies the possibility of psychophysical laws, is the same that founds psychology as a special science irreducible to more fundamental sciences. If psychological kinds do not correspond neatly to physical kinds, then the laws of psychology cannot be reduced to the laws of physics or neuroscience. As Fodor (1974) famously argued in "Special Sciences," the description of what all instances of "believing that it will rain" have in common would be a long and messy disjunction of disparate physical states, not a unified physical kind that could figure in a physical law.

Fodor (1991a) defended the notion that psychology should not be judged by its ability to produce exceptionless, universal laws, like those in physics. Instead, psychological laws are typically ceteris paribus laws, meaning they hold when all other things are equal. These laws express genuine, causally efficacious regularities at the psychological level of description, even if they admit exceptions due to intervening factors at lower levels. We will see this position in more detail in the following item.

Fodor's ideas about the special sciences, ceteris paribus laws in psychology, and a computational theory of the mind (Fodor, 1975) were the most important impulses to the consolidation of the cognitive revolution in the discipline and the computationalist approach. David Marr (1982) was the first to try to offer an entire description of the rules of information-processing governing a psychological (sensorial) capacity, studying vision (under ceteris paribus conditions). Zenon Pylyshyn (1984) tried to extend the ceteris paribus law framework to symbolic cognition. They, along with the computationalist approach, consider the rules of information processing as law-like because (i) they are systematic and predictive, and (ii) failures can be traced to identifiable violations of background assumptions.

Another traditional approach that goes along with Fodor's position assumes that psychology has only statistical regularities, not laws. Paul Meehl (1978) is a classic defender of this approach. Meehl agrees with the traditional obstacles to laws in psychology (such as multiple realization or contextual sensitivity) and concludes that theories from what he calls "soft-psychology" (clinical, counselling, social, personality, community, and school psychology) "lack the cumulative character of scientific knowledge" (p. 806).

Because many uncontrolled factors influence psychological variables, the literal null hypothesis[2] (that there is no difference between X and Y), which is typically tested by such theories, is almost always false. Meehl (1978) asks for "laws" in psychology that would be a high-probability regularity whose stability across diverse samples indicates an underlying causal propensity (e.g., power-function learning curves, robust trait correlations). Confirmation, therefore, employs effect-size estimates and replication, rather than a single significance test.

Meehl (1978) sees the psychological equivalent to "laws" as robust probabilistic regularities rather than strict necessities. To test them, he endorses a Popperian strategy: abandon low-risk null-hypothesis rituals, require theories to forecast precise statistical patterns, and evaluate them by how well those forecasts survive replication under varying conditions. The probabilistic approach to regularities that he championed has been a constant presence throughout the short history of the discipline (Luce, 1986; Glymour, 2001).

Another influential approach to laws in psychology was proposed by Robert Cummins (1983, 2010). In "*'How does it work?' vs. 'What are the laws?' Two conceptions of psychological explanation*,"[3] initially published in 2000, Cummins develops the positions about psychological laws of his 1983 classic *The Nature of Psychological Explanation*.

To Cummins (2010), sciences like psychology and geology are "special" (not fundamental), because the principles of psychology and geology and "other so-called special sciences do not govern nature generally, but only special sorts of systems" (p. 287). Laws of psychology, therefore, are laws that hold only in a special kind of system because of its peculiar constitution and organization. They are not general laws of nature, but rather laws that specify effects of a special kind of "mechanism": human minds.

Cummins (1983) argues that explaining a capacity involves decomposing the system (mechanism) that produced it into sub-capacities, whose joint operation is necessary for the target phenomenon. So, unlike Fodor, he treats these capacity requirements as mechanistic necessities, not laws in the physics sense. He compellingly argues that:

> Once we see that the laws of a special science are specifications of effects, we see why theories in such sciences could not be anything like Newton's Principia. Who would be interested in an axiomatic development of the effects exhibited by the liver or the internal combustion engine? What we want is an explanation of those effects in terms of the constitution and organization of the liver or engine. At the level of fundamental physics, laws are what you get because, at a fundamental level, all you can do is say how things are. (Cummins, 2010, p. 287)

Cummins ideas influenced all mechanistic approach to (negate) laws in psychology, such as the one of Bechtel and Abrahamsen (2005). Mechanism explanations, as we saw, appeal to organized entities and activities, not laws. But the thesis that laws should have a diminished role in psychology may be misleading. The laws of psychology are indeed (several times) the very phenomena to be explained (the effects), as every law that is not fundamental, from every science. However, at some point, even in the mechanistic approach, they will also be the principles that do the explaining. Because at some level, if the explanation goes deep enough, fundamental laws will have to describe how things are, simply. Moreover, there is no reason to discard the possibility that consciousness is something from this fundamental level of nature.

A similar position was defended by Woodward (2000). As we saw here in the chapter on Explanations, Woodward (2003) defends intervention as the mark of causation. He defines "cause" in terms of ideal interventions: X is a cause of Y if and only if there exists some manipulation X that changes Y while holding other variables on their causal paths fixed. To him, explaining is citing such intervention-supporting dependencies.

As we saw, there are some problems with this position, as the confusion between how we identify a cause and how we should define it. However, the most obvious is the problem of what explains the explainer, in other words, what explains that the manipulation of X over Y always (or usually) holds, if not a law. Woodward (2000) is not entirely skeptical about laws of nature (especially in physics), but he does not assume that the regularity usually holds because of one of them. Instead, he introduces his concept of "invariance." To him, "a generalization describing a relationship between two or more variables is invariant if it would continue to hold—would remain stable or unchanged—as various other conditions change" (2000, p. 205).

One fundamental difference between the concept of invariance and the concept of law is that invariance is relative to a specific context. To Woodward, typically, a relation will be invariant concerning a specific range of changes but not concerning others. Another crucial difference is that stability, not exceptionless (reliability), is the criterion of invariance. A third is that invariances come in a continuum, as Woodward (2003) suggests, from a maximum invariance that holds under every possible intervention (equivalent to sufficient laws) to very narrow invariances, which hold only within a narrow band of manipulations.

This last kind of invariance is the only one available in special sciences like psychology, which can only expect law-like generalizations, which hold after at least some ideal interventions on the variables it cites. While physical laws tend to survive a vast class of manipulations, psychological generalizations survive only a narrow band (Woodward, 2000, 2003).

The reasons for the inexistence of laws (maximum invariances) in psychology, according to Woodward (2000), are the usual. First, psychological phenomena are too context-sensitive; human behavior frequently depends on background factors (attention, motivation, emotion, information) that are themselves hard to manipulate. Second, the problem of multiple realizability, which makes psychophysical laws impossible, since diverse neural structures can realize the same function, so a strict correspondence between them cannot be achieved.[4] Third, the difficulty of isolating variables, since in this highly complex system, an intervention that would be straightforward in, say, physiology (lesioning a single nucleus) may disrupt an entire cognitive network.

Nevertheless, despite these obstacles, Woodward (2000) concedes some modest invariances, from the lowest level, in psychology. They qualify as laws in the pragmatic sense that Woodward accepts them, because they enable scientists to manipulate, predict, or control certain phenomena.

Behind all the positions pro psychological laws (or invariances), there are not many pragmatic differences. In comparison to Woodward's position, Fodor's ceteris-paribus laws are acceptable if they pick out variables whose manipulation is known to disrupt the regularity. Regarding explanation, Cummins's functional analysis is explanatory when it identifies sub-capacities whose manipulation (lesion, overload, suppression) would prevent the capacity. I argue here that psychology's difficulties with laws comes from the fact that it has just searched for the wrong laws all along: laws like physical laws. Moreover, it has been ongoing for so long that many researchers have abandoned this search.

Sufficiency laws in psychology have been condemned since the very beginning. Their failure has the usual suspects. There are too many variables, too many complexities, too much information and contexts to consider, too little ecological validity[5] in experimental research. Therefore, psychology has moved long ago to probabilistic generalizations based on statistical analysis and to mechanistic models. Although they have merit and valuable applications (such as in cognitive-behavioral therapy, cognitive assessments, and neuropsychological tests), progress toward a unified theory of the human psyche has been almost null. Furthermore, these models and generalizations offer little predictive and explanatory power, if any, for individual behavior. They are so imprecise that they can barely be considered law-like.

So, can we talk about "laws" in psychology? I defend that, more than mere invariances, psychology has two perfectly operational types of laws. I am about to argue here that those two contenders to probabilistic generalizations and failed sufficient laws have already entered the stage of psychology, even if psychologists are not entirely conscious of them. They work very well in the discipline, and do so, leaving us the possibility of some autonomy in the face of the social, biological, and psychological conditions to which we are subjected. The first of them is the comparative ceteris paribus law. The second, the PL.

6.2.1 Ceteris Paribus Laws in Psychology

Ultimately, no sufficiency laws in psychology can predict what any individual's behavior will be like, not even one, despite more than 140 years of experimental research in the discipline. That should say something eloquent to those living in the world of facts rather than in their metaphysical beliefs. If sufficiency laws require ceteris paribus clauses even in physics (Cartwright, 1983), we should expect that generalizations and candidates for sufficiency laws in psychology should be ceteris paribus. Let us examine this problem in psychology without skipping steps.

6.2.1.1 Hedged Laws

Stephen Schiffer (1991) also thinks there is no such thing as psychological laws. Roughly, he argues that if there are psychological laws, they must be hedged laws. A hedged law includes additional conditions or clauses alongside the primary condition for its (sufficiently) predicted outcome to occur. However, Schiffer says there are no hedged psychological laws because psychological states are highly complex and variable. This makes it impossible to formulate generalizations that hold across different contexts and individuals. Therefore, there are no psychological laws at all. In other words, according to Schiffer, hedged psychological "laws" are not laws because they cannot specify a genuine clause that, together with a specific condition, always causes a kind of result. They would not have empirical content because they would not have the capacity to make precise predictions that can be empirically corroborated or falsified based on observable and measurable outcomes.

When Schiffer (1991) talks about hedged laws in psychology, he talks about ceteris paribus laws. There are, however, at least four problems in his argumentation:

There are hedged laws that are no ceteris paribus, and they work very differently from the latter.

There are hedged laws that are no ceteris paribus, and they are perfectly falsifiable.

There are two kinds of ceteris paribus laws, and they work, also, very differently from each other.

It is false that all kinds of ceteris paribus laws have no empirical content.

Let us not put the cart before the horse. To better understand ceteris paribus laws, we should first specify what a hedged law is. I previously defined a scientific law—a proposition that represents a law of nature—as a universal quantification of a material implication. Its form is $\forall x(S(x) \rightarrow R(x))$; where $S(x)$ is a predicate that describe "specific" conditions or events that obtain at time t1, and $R(x)$ is a predicate describing the "result," the outcome that necessarily follows at time t2 when $S(x)$ are met. We could read this formula as "For all x, if x satisfies the specific condition S in t1, then x necessarily has the outcome R at t2."

In turn, a hedged law is a kind of law that introduces an additional clause on this formula, a "completer" (Fodor, 1991a). This clause captures background or contextual factors—C(x)—that, together with S(x), are jointly sufficient for the result R(x). Therefore, a hedged law can be represented as $\forall x\ ((C(x) \wedge S(x)) \rightarrow R(x))$. Importantly, both C(x) and S(x) refer to conditions restricted in time, which may include events or states whose realization is presupposed by the law. This relationship guarantees that if conditions (completer) C and event S occur, event R must occur. The formalism abstracts from how these conditions come about, but it does not deny that they may result from prior causes or processes.

Fodor (1991a) specified the necessary and together sufficient conditions for C to be a completer. (i) $C(x) \wedge S(x)$ is strictly sufficient for R(x). (ii) It is not the case that C(x) alone is sufficient for R(x). (iii) It is not the case that S(x) alone is sufficient for R(x).

Let us consider a straightforward psychological example from developmental psychology. It is well known that children acquire language between the ages of two and five, provided that they are exposed to language daily and have no neurocognitive disorder or impairment (Hoff, 2006; Kuhl, 2010). There are only two conditions that must be satisfied in its completer. Let us formalize it to see better:

$$\forall x \, ((H(x) \land (E(x) \land \neg I(x))) \rightarrow L(x)).$$

Where we read, “For all x, if x is a human baby (H), and has daily basis exposition to language (E) and does not have a neurocognitive impairment (I), then x will acquire language (L) until the age of five. $E(x) \land \neg I(x)$ is the completer, the clause that must be satisfied for H to instantiate L.

Despite the usual difficulties in operationally defining these concepts, we have here a perfectly reasonable hedged psychology law that is easily falsifiable. It can be falsified by a single instance of a human child being exposed to language daily and not experiencing any neurocognitive impairment (such as cerebral palsy), yet not acquiring language by age five.

Moreover, it is also not a ceteris paribus law because of the nature of its clause. The completer does not demand that everything else remain equal or absent to fulfill the prediction. It simply requires the absence of impairment and daily exposure to language.

However, what makes simple hedged laws distinct from ceteris paribus laws? Simple hedged laws are combinatorial laws (Rupert, 2008). They mix two different orders of causation in one specification: sufficient and necessary. In other words, simple hedged laws mix a comparative ceteris paribus law (CCPL) with one or more PLs. To demonstrate this, I should first explain the nature of a CCPL.

6.2.1.2 Two Kinds of Ceteris Paribus Laws

Hedged laws can have ceteris paribus clauses. Authors who defend that there can be genuine laws with these clauses (Fodor, 1991a; Carrier, 1998; Gadenne, 2006; Rupert, 2008) say that they can be of two types: “if S, then R, other things being equal,” which form comparative CPLs (CCPLs), or “if S, then R, excluding all interfering factors,” which form exclusive CPLs.

Nevertheless, several other theorists deny the possibility of CPLs, considering them, strictly speaking, false (Cartwright, 1983; Schiffer, 1991; Woodward, 2002). CPLs would not qualify as laws because they would not be universal (unlimited in extent). There are situations where they simply do not work. A CP clause would protect the law from its counterfactuals. In other words, all possible results in the phenomenon governed by it would ultimately be covered by the CPL: it would be unfalsifiable and untestable.

First, I would like to reclaim common ground and rule out exclusive ceteris paribus laws as scientific law candidates. The format “if S, then R, excluding all interfering factors” is not empirically testable in psychology or any science. To limit this reflection to psychology, try to imagine how we would isolate an actual human being from all his background, biological processes, environmental information, and cognitive processes, only to test the effect of alcohol on his reflexes. It is not that we cannot make it; it is that we cannot even imagine it. It is impossible. Therefore, it cannot be tested, and therefore, it cannot be scientific.

Therefore, we should focus our investigation on CCPLs. Let us begin by specifying it. As a type of hedged law, a ceteris paribus law (CCPL) includes a completer alongside the specific condition S; in this case, the completer specifies that all conditions other than S must remain constant. That is, R necessarily follows from S only when no other relevant conditions change. We have therefore $\forall x\ ((CP(x) \land S(x)) \rightarrow R(x))$. Where CP is the completer (other things being equal: Ceteris Paribus), S is the description of an event specified in the antecedent, and R is the result. We can read this: "For all instances x, if event S occurs and all conditions except S remain constant, then result R follows."

As a type of hedged law, a ceteris paribus law (CCPL) includes a completer alongside the specific condition S; in this case, the completer specifies that all conditions other than S must remain constant. That is, R necessarily follows from S only when no other relevant conditions change.

6.2.1.3 Ceteris Paribus Laws as Expressions of Causal Pressures

The central problem imposes itself from the beginning: What is the empirical content of a CCPL? How could we test it? The opponents (like Cartwright, 1983 and Schiffer, 1991) accuse CCPLs of impossibility in testing because the ceteris paribus clause would be too inclusive, excusing exceptions. I should justify why I think the opposite, that CCPLs can handle exceptions and have empirical testability.

To begin, a CCPL has an unlimited domain; the postulation is universal: "every time, everywhere, as other factors remain constant, if S, then R." The problem, therefore, is not that it is not universal and has not unlimited scope. The problem is the empirical content of such affirmation. In other words, the problem is how this prediction can be falsified; as to any individual instantiation of the law, we can say that it did not hold because some factor may not have remained constant.

Let us consider a CCPL that states, "other things being equal, an increase in the blood alcohol level decreases the driver's reflexes." In this case, we would not be able to explain why a specific drunk man caught in a traffic stop does not show reduced reflexes. However, this event would not falsify the law, as we could always assume that intervening factors did not remain constant (we could speculate that he was simultaneously under the effect of a drug that increases the speed of reflexes).

That is why Fodor (1991a, p. 49) argues that CCPLs only have the force of law in situations defined by the CP clause or "completer" in his classification. Nevertheless, Carrier (1998) contends that a ceteris paribus law does not lose its status as a law with unrestricted spatiotemporal validity simply because, in a particular instance, the expected outcome (such as "if more alcohol, then less reflex") does not occur. Thinking this way is a distortion that arises from representing a law as a logical relationship rather than a causal relationship.

Rupert (2008) distinguishes strict laws from ceteris paribus laws through the concepts of non-combinatorial laws and combinatorial laws. The properties of a CCPL's antecedent (e.g., alcohol) cause their standard effects (decrease the speed of reflexes) every time they are instantiated. However, sometimes, these causal influences combined with others (e.g., a drug to increase reflexes) in a combinatorial law (of the functioning of reflexes) do not lead the system to behave as we expect (decrease reflexes), leading many to believe that this is an exception to the law. It is not. In typical real systems, multiple non-combinatorial laws and multiple causal influences operate together to determine the system's behavior. This composition of forces is what Rupert calls "combinatorial law."

Let us continue with the previous example: "Other things being equal, an increase in the blood alcohol level decreases the driver's reflexes." In some events, it is the case that a drunk man D., caught in a traffic stop, does not show reduced reflexes. The adversaries of ceteris paribus laws, as actual laws, would say that this event does not falsify the law, as we could always speculate that some intervening factors did not remain constant (he consumed a drug that simultaneously increased his reflexes). And they would be totally right.

However, even if we cannot "exclude all other factors" as the exclusive CPL asks, we can still "hold all other factors" constant for experimental testing purposes with relative ease. The critical point here is that we cannot do this for a single case. We can only achieve this for a sample in an experiment with a random distribution of subjects between the test and the control group.

In theory and practice, a random assignment of subjects to experimental and control groups in an experiment with a sufficiently large sample equalizes the known variables and any hidden or unmeasured variable across the groups. The random distribution assumes that every individual with any level of an unmeasured variable in the population has an equal chance of being assigned to any of the groups. Therefore, its sum will have approximately equal levels across the two groups (all other things being equal). This method dilutes the effect of individual differences and uncontrolled variables, distributing these elements as evenly as possible across all groups.

From the fact that probabilistically, a random assignment makes any observed differences in outcomes across groups more confidently attributed to the variables being manipulated in the experiment, a CCPL is tested in experiments, albeit not in individual cases. Of course, this confidence depends on having a sufficiently large sample size. The larger the sample size, the higher the reliability of the equal distribution assumption and the lower the influence of strange variables. This is why, to determine the adequate size of a sample, a psychologist must first define what she considers statistically significant.

Therefore, the CCPL "holding other conditions constant, an increase in the level of alcohol in the blood decreases the driver's reflexes" is testable by a classic experiment with a randomly divided sample, where we can measure the value of R with the certainty that the other factors remained approximately constant between the two groups, as they are large enough and were distributed randomly. Remembering that, however, as we saw above, we cannot predict the individual case; we can only predict the behavior of samples.

This trait leads to my second line of defense here. Being able to predict the behavior of samples, but not that of individuals, indicates that a CCPL accurately captures not a logical relationship but a constant causal pressure in nature. Take a classic example of a candidate law in psychology: Thorndike's law of effect, better known as Skinner's formulation of operant conditioning. As the law cannot express a precise mathematical relationship, it only states that if a behavior is reinforced (followed by a reward), its frequency will increase; if it is punished (followed by an aversive stimulus or the loss of a rewarding stimulus), its frequency will decrease. When faced with a bar that releases water into its cage, we cannot sufficiently predict how the rat will behave in the next five seconds. However, we can say that since the behavior of pressing the bar has already been reinforced, the probability that it will be pressed over the next five seconds has increased (relative to the base rate of the behavior, as measured at the beginning of the individual session). A single measurement may not be able to measure this, but twenty will not fail. Likewise, we cannot predict that a particular driver's reflex will decrease when faced with a dose of alcohol; instead, the probability of this happening increases. However, the average reaction time will inevitably drop in the experimental group if we apply the test to a randomly distributed sample.

What a CCPL can capture, therefore, is the causal pressure of a given contributory condition for a given behavior. Furthermore, that is precisely what psychological life is like. The lower the blood water level, the greater the pressure to drink; the lower the blood glucose level, the greater the pressure to eat; the acquisition of a particular belief, such as that a building is about to collapse, increases fear, which increases the pressure to escape. However, in none of these cases do these pressures (which can be perfectly measured and even estimated by CCPLs) sufficiently cause the behavior they pressure for.

In summary, comparative ceteris paribus laws (CCPLs) are falsifiable within a confidence interval, as a sufficiently large sample with random distribution of subjects can ensure that all other relevant factors remain approximately constant. Exclusive ceteris paribus laws, by contrast, can never fully exclude all interfering factors in any test and are therefore not empirically testable. CCPLs, then, are falsifiable (though not by single observations), support counterfactual reasoning, and enable sample-based behavioral predictions. They can be used in explanations to justify why, in a particular case, a pressure toward a certain behavior has been raised, though never to claim that such pressure was itself sufficient for the behavior. CCPLs are, at minimum, well-supported empirical generalizations, and likely express natural regularities—specifically, causal pressures. My claim here is epistemic, though it is plausible that these generalizations also reflect an underlying physical reality.

6.2.1.4 The Need for a Better Formalization

How do we capture this intuition in a logical formula that facilitates a more precise prediction and, therefore, a more precise eventual falsification of this kind of law? The form $\forall x\ ((CP(x) \wedge S(x)) \rightarrow R(x))$ is a correct description of a ceteris paribus law, but it has limitations in capturing a causal pressure.

The first limitation is that S and R would always have to be defined as an increase or decrease of some variable. The second, derived from the first, is that using them in the same formal explanation with PLs would be difficult. The third is that it would be less explicit to sustain predictions that could be falsified.

To solve this and make it more explicit formally and mathematically, we should mix non-standard logic inside object-language, using the symbol $\uparrow\Delta$ to express that an increase in the variable S will increase the probability of the result R in a CCPL formula as follows:

$$\forall x[CP(x) \wedge \uparrow\Delta S(x) \rightarrow \uparrow Prob(R(x))]$$

This can be read as follows: "For all instances x, if x is under the ceteris paribus condition (where other things remain equal) and there is an increase in S (the specific condition/event S changes positively), then the probability of R (the result/event R) increases."

Given the example "Other things being equal, an increase in the blood alcohol level decreases the driver's reflexes," the CCPL can be formalized as:

$$\forall x(CP(x) \wedge \uparrow\Delta BloodAlcoholLevel(x) \rightarrow \uparrow Prob(DecreasedReflexes(x)))$$

This formalization expresses how an increase in a specific condition (alcohol level) will increase the causal pressure in R (decreased reaction time), reflected in the increase of the probability of the result R. This is why we should not use a formalization with direct changes, this would suggest a more direct and deterministic relationship where a change in S directly leads to a change in R in that exact proportion.

On the other hand, a probability formalization (of the result variable) is a more precise representation of contexts where the effect is not guaranteed but is influenced by other factors, such as in psychology, medicine, and social sciences. This approach allows us to capture the causal pressure and the probabilistic nature of the relationship between the "realizer" S and the increase or decrease in the "result" R.

Comparative CPLs are everywhere in psychology, from Skinner's operant conditioning to the increasing effect of physical arousal on emotions, from the effectiveness of repetition in forming long-term memories to the priming effect on implicit memories. The primary causal relationships established by psychology can be expressed in this type of law but not in strict sufficiency laws or even probabilistic laws that express genuine probabilistic relationships, as in quantum physics.

In addition to being more realistic and honest, comparative CPLs are predictive and useful. We can estimate the water level in the blood at which 1 percent of people start to become delirious, and based on this, safe consumption rates on space missions, for example. We can increase the incidence of safe behaviors in children via praise or improve the chances of getting out of depression via an antidepressant. We can point to study methods that facilitate content retention, but we cannot determine the specific behaviors that result from these methods. This inability occurs, certainly, due to the enormous complexity of the causal pressures affecting the most complex of objects known in the universe, and possibly, because human consciousnesses are initiators of new causal chains.

6.2.1.5 Examples of Comparative Ceteris Paribus Laws in Psychology

CCPLs in psychology are particularly useful where isolating variables is extremely challenging due to the complexity of human behavior. With experimental tests of CCPLs, this difficulty is largely mitigated by a large sample size and a random distribution. Several psychological principles that appeared limited, permissive, and unreliable reveal themselves under this light as laws exercised universally and reliably, though permissively. The most classic of them can be illuminated by a reinterpretation as causal pressures, captured and represented by CCPLs, which offer a rational framework to their apparently nondeterministic character. For the sake of simplicity, I will represent them here as regular CCPLs, expressing the probabilistic effect on results in the description of the variables.

6.2.1.5.1 Classic Conditioning

Pavlov's classic principle (1927/2010) can be considered a CCPL. It describes how an organism can learn to associate a previously neutral stimulus (a stimulus that by default elicits no particular response) by repeatedly pairing (RP) it with an unconditioned stimulus (a stimulus that innately elicits an unconditioned response). Each pairing of these kinds of stimulus raises the force of the association, until the neutral stimulus eventually becomes a conditioned stimulus (CR) that elicits a conditioned response similar to the original unconditioned stimulus.

We can formalize it as:

$$\forall x\ ((CP(x) \wedge \uparrow\Delta RP(x)) \rightarrow \uparrow Prob(CR(x)))$$

Where:

x: an animal organism

CP(x): "Other things being equal."

$\uparrow\Delta$RP(x): Repeated Pairings of the neutral stimulus with the unconditioned stimulus increase.

$\uparrow$Prob(CR(x)): The frequency of the conditioned response (CR) increases.

We can read this as "For any organism x, if all other conditions remain constant, and the frequency of the pairing of a neutral stimulus with an unconditioned stimulus increases, then the probability of the conditioned response will increase." This probability can be read as the expected frequency in a large sample.

The terms unconditioned and conditioned here reflect whether a response occurs naturally (without learning) or as a result of learning. The UR is called "unconditioned" because it happens reflexively or automatically in response to the US, without prior learning. When the NS is paired with the US often enough, the organism begins to respond to the NS alone—a learned or "conditioned" response.

For example, in Pavlov's famous experiment, a bell (NS) was repeatedly paired with food (US), which naturally caused salivation (UR) in dogs. Over time, the bell alone (now a CS) came to trigger salivation (CR), even when no food was present.

Classical conditioning is a CCPL because its variability (permissibility) is due to differences in organisms, different cognitive activities in different species or specimens, prior associations of the neutral stimulus in some specimens, concurrent association with the neutral stimulus during the pairing, inconsistent timing of pairings, and the list goes on infinitely. How do we control it to test? Equal independent variable manipulation patterns, a large sample of individuals, and random distribution between the groups. We cannot predict the strength of conditioning in any individual. However, we will be able to measure it approximately in a large sample, confirming securely the causal pressure caused by the pairing.

While classical conditioning is a powerful learning mechanism, it is not deterministic. However, this does not mean that it is not a law. It is just a law immersed in a sea of competing causal pressures in the combinatorial law of behavior. It always exercises its own pressure, but is not always decisive in the behavioral outcome. Even within the law, the strength of the CR can vary depending on factors like the number of pairings or the intensity of the United States.

Recognizing classical conditioning as a CCPL not only deepens our understanding of its apparently probabilistic functioning and conditional nature but also reveals the logical form through which it exerts its causal power. This has practical implications in various fields. For instance, in marketing, an advertisement might repeatedly pair their product (NS) with images of happy people, beautiful scenery, or upbeat music (US). Over time, consumers will start associating the product (now a CS) with these positive emotions (CR), leading to a more favorable attitude toward the product and an increased likelihood of purchase. Although none can necessitate purchase behavior, the rise of favorable attitudes and the likelihood of purchase are sure and measurable. Understanding that psychology can only predict the increase or decrease in the pressure of behavior does not serve only advertisers, but especially the general public, in resisting manipulation consciously.

6.2.1.5.2 Operant Conditioning

Operant Conditioning is the modern version of Thorndike's Law of Effect, which was refined by Skinner (1938), who consistently resisted presenting it as a law. Operant conditioning is one of the simplest and intuitive psychological principles. It says that if a satisfying consequence follows a behavior in a particular situation, the frequency of this particular behavior will rise in the same situation (the association between the situation and the response is strengthened). If an unsatisfying consequence follows, its frequency will fall. All we have to do to turn it into a proper CCPL is to add the clause "other things being equal." We can formalize it as:

Reinforcement:

$$\forall x\ ((CP(x) \land \uparrow\Delta R(x)) \rightarrow \uparrow Prob(B(x)))$$

Punishment:

$$\forall x\,((CP(x) \wedge \uparrow\Delta P(x)) \rightarrow \downarrow Prob(B(x)))$$

Where:

x: an organism in a specific environment

CP: "Other things being equal" or "All other conditions affecting behavior remain constant."

R: reinforcing stimulus rate (reward).

P: punishing stimulus rate (punishment).

The reinforcement formula can be read as "For any organism, if all other conditions remain constant and the rate of rewards following its behavior increases, then the frequency of that behavior will increase." For the punishment, we can read, "for any organism, if all other conditions remain constant and the rate of punishments following its behavior increases, then the frequency of that behavior will decrease."

Skinner (1938) did not explicitly formulate operant conditioning as a law. As a conscientious operationalist, he could never concede the status of law to such a permissive principle. He was satisfied (maybe) in describing it as a set of principles (or a "function") that explain how behavior is shaped by its consequences. Also, the force of a reinforcement stimulus decreases in time. The formalization of operant conditioning as CCPLs not only highlights the conditional nature of this law but also makes clear that their effect is unlimited (happens in all instances of animals and behaviors) and reliable (happens always) despite the final result depending on the "combinatorial law" of behavior. This helps to clarify its nature of law but differentiates it from deterministic laws that predict outcomes with absolute invariability (non-permissively).

Few laws are more important to human beings than operant conditioning. From the education of our children to formal learning, parents use positive reinforcement, such as praise and smiles, and teachers use it to encourage desired behaviors like studying and completing assignments. Therapists employ operant conditioning in behavioral interventions for addiction, anxiety, and autism. Within the workplace, managers utilize reinforcement strategies to enhance productivity. Understanding the character of causal pressure of this clear and basic law of psychology helps us to resist the tendency to repeat from the opening of the refrigerator to superstitious behavior. Which laws of biology or economics do you know that are so widespread?

6.2.1.5.3 Dunning-Kruger Effect

Let us get synthetic and only specify three more CCPLs of psychology. For the Dunning-Kruger Effect, an important law in times of social media, the more incompetent a person is in a particular area, the more they are likely to overestimate their competence in that area (other things being equal). This cognitive bias shows how people's lack of knowledge can lead to overconfidence. It is an important source of heuristics to put under suspicion when we are overconfident in our opinions about an area that we do not know well. We can state it as:

$$\forall x\ ((CP(x) \wedge \uparrow\Delta I(x)) \rightarrow \uparrow Prob(O(x)))$$

Where:

x: an individual

CP(x): "Other things being equal" (especially other factors influencing self-assessment as personality traits, emotional state, substance abuse . . .).

I: Incompetence score in the specific domain.

O: Overestimation of one's ability in the specific domain.

That we can read as "For any individual, if all other factors remain constant, the higher a person's incompetence score in a specific area, the higher the probability of overestimating one's ability."

6.2.1.5.4 Cognitive Dissonance

One of psychology's most classic and well-known principles should be seen as a CCPL. Festinger's Theory says that if a person holds two cognitively inconsistent beliefs or their behavior contradicts their beliefs, then they will experience discomfort (dissonance) and will be motivated to reduce this dissonance (other things being equal). We can represent it as:

$$\forall x\ ((CP(x) \wedge \uparrow\Delta D(x)) \rightarrow \uparrow Prob(M(x)))$$

Where:

x: a person

CP(x): all other factors influencing motivation remain constant.

D(x): level of cognitive dissonance

M(x): Motivation to reduce inconsistency

We can read it as "For all persons, if a person's level of cognitive dissonance increases, and all other factors remain constant, then the probability of a person's motivation to reduce inconsistency increases."

To diminish or eliminate this discomfort, individuals feel the motivation to change either their beliefs or their behaviors to align them. An example is the cognitive dissonance experienced by a man who believes that smoking is harmful, but continues to do it. To reduce this discomfort (dissonance), they might change their belief (convince himself smoking is not too harmful) or change his behavior (quit smoking).

6.2.1.5.5 Recency Effect on Memory

Let us conclude this long defense of CCPLs by examining a case where one of them was falsified. It shows how they possess empirical content and are easily falsifiable. The recency effect, in its original form, predicts that items from a list presented later are more likely to be remembered (all other factors equal). We can represent this as:

$$\forall x\ ((CP(x) \wedge \uparrow\Delta P(x)) \rightarrow \uparrow Prob(R(x)))$$

Where:
x: an item of information
CP: all other factors influencing recall remain constant.
P: Proximity of an item to list-end.
R: the likelihood of Remembering the item.

We can read this as: "For all items of information, nearer an item is to the end of the list, the higher the probability it will be recalled, all other factors remaining constant."

This law is a CCPL because it does not always hold true. Various factors can alter the expected result, such as the length of the list, the time between presentation and recall, individual differences in memory capacity, the meaning of items (it is preferable to do those tests with non-sense syllables), the emphasis of the presentation of the items by a narrator, and the list goes on. Some of these items we can control in an experimental situation; others, only through a random distribution in a significant sample size.

For instance, with longer lists, we also observe a primacy effect, where items at the beginning are better recalled due to factors like repetition and increased attention. This finding falsified the simple recency effect, leading to the development of the more comprehensive serial position effect (Murdock, 1962), which accounts for both primacy and recency.

The recency effect is attributed to the items still active in short-term memory during recall. However, there are other processes in action. The presence of the primacy effect, influenced by repetition (a process of long-term memory consolidation), reveals the limitations of the original recency effect law. This example shows how CCPLs, while not absolute, are valuable tools in psychological research, predictive and falsifiable. They allow us to isolate and examine causal relationships while acknowledging the complexity of variables in real-world scenarios. With this example, I rest my case for CCPLs in psychology.

6.2.2 Possibilist Laws in Psychology

Let us quickly re-present, for didactic exposition purposes, what would be a PL in psychology. A long-standing neurocognitive claim is that only in the presence of preserved hippocampi is it possible to form new long-term memories (Scoville & Milner, 1957). If we have or do not have a new memory in the presence of preserved hippocampi, the sentence is corroborated. However, what an LP predicts can be tested is not this possibility but the impossibility of forming a new long-term memory with compromised hippocampi.

The form of a PL, therefore, should be $\forall x\ (\neg Hx \rightarrow \neg\Diamond Mx)$, where H is "preserved hippocampus," ¬H we can read as "compromised hippocampus" (non-preserved hippocampus), and M is "new long-term memory."

The structure of a prediction in PL is that an effect will not happen without its necessary condition. If it happens, the law is falsified. The structure of the falsification, therefore, is (a) PL': "If there is no H (preserved hippocampus), there is no M (new long-term memory)"; (b) It is the case of having M (new memory) without H (in a person with the hippocampus divided into both hemispheres); (c) Therefore, PL' is falsified.

This negative necessity claim [$\forall x\,(\neg Nx \rightarrow \neg\Diamond Rx)$] sits between the over-permissive ceteris paribus hedges of classical special-science laws and the over-demanding ideal of strict sufficiency law. It inherits or adapts the main insights of the psychological laws' approaches surveyed at the beginning of this section.

Davidson (1970) have argued that psychology cannot yield genuine laws because of the "anomalism of the mental." The meanings and norms involved in mental life would resist generalization under universal laws. But PLs, as defined here, do not require universal application. They capture necessary conditions under which behaviors or processes become possible or impossible, rather than offering sufficient conditions for their occurrence. Therefore, they can be epistemically valid even if the total context is open-ended and normatively structured.

While traditional Fodor's CPLs leave the list of "interfering factors" implicit, a PL spells out necessary conditions hidden in the enabler set. If someone thinks a factor is a necessary condition N, the law predicts that removing it eliminates R. This is an empirical test, not a rhetorical escape clause.

As for Cummins' (2010) functional analysis, we can say that each sub-capacity in a functional decomposition is necessary for the overall capacity. The possibilist calculus turns that verbal necessity into a modal law: "without lexical retrieval N, sentence production R cannot occur," making Cummins-type mechanisms and explanations experimentally falsifiable. Cummins objection that psychological explanation proceeds not through laws but via functional decomposition does not eliminate modal constraints. There are conditions without which the function cannot operate. PLs formalize these necessary boundaries in a way compatible with functional modeling.

Woodward's invariance account also fits nicely in a possibilist frame. A law must remain stable under interventions on background variables. PLs set an upper bound on invariance: no intervention can restore R once N is disabled. It satisfies Woodward's criterion in the strongest possible sense, since the invariance collapses to zero when N is absent.

Even the basic intuition of the sceptics of psychological laws is contemplated within a possibilist framework. The strict sufficiency of some X causing some Y in psychology remains unattainable for PLs. They do not claim that N guarantees R, but that the absence of N prevents R. Sufficiently. Strictly.

Let us suppose an example. Take working-memory interference. In many dual-task experiments, complex reasoning (R) fails when the phonological loop (N) is saturated (Gathercole et al., 2004). A possibilist formulation could be "if there is no free phonological loop, complex reasoning is not possible. We can test it by overloading the phonological loop while offering subjects money to perform complex reasoning tasks. If we observe that reasoning collapses even with motivational boost intervention, we can say that the possibilist prediction is corroborated, passing through Woodward-style invariance.

Here we have explicit falsifiability, tolerance of heterogeneity, and compatibility with mechanistic explanations. Explicit falsifiability because, unlike bare CP clauses, the impossibility claim is defeated by one verified case of R co-occurring with ¬N. Tolerance of heterogeneity because different neural or computational implementations of N do not threaten the law. Finally, compatibility with mechanistic explanations is evident because identifying N typically involves uncovering a mechanism (loop, receptor, control module). Taken together, PLs retain the domain-specific explanatory power of Fodor's CPLs while addressing long-standing objections to psychological laws regarding vacuity, invariance, and multiply realized kinds.

PLs are also everywhere in psychology. Every subdiscipline of psychology has nontrivial examples of PL candidates. It is important to emphasize here that what I am asserting is that these laws are actually present in psychological life and, therefore, are present, implicitly, in theories and models throughout the history of psychology, although they have not been formalized in the discipline. Furthermore, I think they haven't been formalized because we simply haven't been looking for them in the last few centuries.

Certainly, there are PLs in psychology. But what kind of laws are they, exactly? More specifically, what is the modal strength of these laws? Are the possibilities and impossibilities they describe epistemic or (meta)physical? Are they biological or psychological? As with CCPLs, my claim is that PLs *certainly* reflect practical constraints and *probably* express metaphysical limits. It is an empirical fact that certain behaviors depend on necessary conditions, and it is plausible that our knowledge of these conditions (scientific laws) ultimately derives from physical boundaries (laws of nature). For example, consider the fact that Steve cannot walk due to the absence of neuronal connections between his brain and his legs. In this case, the impossibility is biological. However, we can also conceive of purely psychological impossibilities, such as learning multiplication before mastering addition.

Let us begin by illustrating this with a PL we have already encountered, perhaps without explicitly recognizing it. Let us consider the case of hedged laws.

6.2.2.1 Hedged and Possibilist Laws

In the last section, I affirmed that simple hedged laws (non ceteris paribus) are combinatorial laws that mix a comparative ceteris paribus law (CCPL) with one or more PLs. Now, I must justify that claim.

In Fodor's (1991a) formulation, a hedged law has a clause (completer) that, together with the antecedent of a material condition, instantiates the consequent. I represented it as $\forall x\ ((C(x) \wedge S(x)) \rightarrow R(x))$. The necessary and sufficient conditions for C to be a completer. (i) $C(x) \wedge S(x)$ is strictly sufficient for $R(x)$. (ii) It is not the case that $C(x)$ alone is sufficient for $R(x)$. (iii) It is not the case that $S(x)$ alone is sufficient for $R(x)$.

At first glance, someone can think that C (the completer) is a necessary condition to R and that $\forall x\ (\neg Cx \rightarrow \neg\Diamond Rx)$ is accurate, a PL hidden into a hedged law; however, this conclusion is not necessarily true because it is derived from another version of the antecedent fallacy.

Let us return to our example of development psychology (Hoff, 2006; Kuhl, 2010). Any children acquire language between the ages of two and five, provided that they are exposed to language daily (E) and have no neurocognitive disorder or impairment (I). The completer has two conditions to formalize $(E(x) \wedge \neg I(x))$.[6]

It seems clear that adequate exposure to language in the developmental window is a necessary condition for a human baby to acquire language by five, but not for every being to acquire language by five (a baby alien would not need it). That may be funny, but it is serious: We cannot make spurious derivations nor misunderstand the causal nature of the law by a logical one.

Another complication is that the conditions in the completer may be permissive and not necessary for language acquisition. For example, not having cognitive impairment or disorder could be permissive of preserving the brain's language functions. We could consider cognitive impairments, such as dyslexia, agnosia, or color blindness, that do not interfere with the acquisition of spoken language. In this case, the absence of cognitive impairment implicates the preserved language functions of the brain, which are the actual necessary conditions for every human baby (not a baby rabbit) to acquire language by five.

These complexities are not merely obstacles; instead, they are integral to scientific work. We may initially misidentify a necessary condition or a sufficient condition. However, the essence of scientific work lies in the continuous refinement of our understanding, replacing supposed necessary conditions (and actual permissive ones) with more accurate candidates to it, as we do refine our understanding of sufficient conditions through more elaborate hedged laws (or accepting them as a causal pressure represented by a CCPL).

6.2.2.2 Discrete Necessary Conditions in Psychology

Let us examine two important concepts for the functioning of PLs in psychology. They are the concepts of "discrete necessary condition" and "continuous necessary condition." Gary Goertz (2003) defines a "discrete necessary condition" as the condition that must be present for an outcome to occur and is typically binary. Using examples from social science, he argues that a country must have a constitution (a discrete condition) to be a constitutional republic. In this context, the necessary condition is either present or not, with no middle ground. Continuous necessary conditions, on the other hand, involve variables that exist on a spectrum rather than a binary state. Degrees rather than absolutes characterize these conditions. For instance, economic development can be considered a continuous variable that influences a country's likelihood of adopting democratic reforms.

My disagreement with Goertz is that, for me, any "continuous necessary condition" either has a threshold beyond which it becomes discrete or is a contributory condition. There are only all-or-nothing necessary conditions. That is not to say that Goertz does not have a solution for it. It is that he considers it practical under certain circumstances. I consider it mandatory to transform them into PLs.

Goertz (2003) argues that it is practical to treat a continuous necessary condition as discrete under certain circumstances. This is particularly true when there is a clear threshold beyond which the condition can be considered effectively binary. For example, suppose we determine that a GDP per capita of $10,000 is the threshold beyond which a country will maintain a stable democracy. This continuous variable (GDP per capita) can be treated as a discrete condition (above or below $10,000).

For Goertz (2003), the decision to convert a continuous necessary condition into a discrete one should be based on empirical evidence and theoretical justification. First, it should have a clear point that distinguishes between the presence and absence of the result. Second, this conversion must have practical utility, improving our ability to explain and manipulate data. Third, the consequent discrete necessary condition must be reliable.

Let us analyze this problem using two (already used) examples of possible law candidates in psychology. Finding actual discrete necessary conditions in this discipline is not so easy. For example, what constitutes a "compromised hippocampus" that makes new long-term memories impossible?

The hippocampus is not a unitary structure. There are two hippocampi, one in each hemisphere of the brain. They are connected by nerve fibers forming the hippocampal commissure structure. A "compromised hippocampus" would refer to a hippocampus that has been significantly damaged or disconnected from other brain regions. But what does "significantly" mean?

In some cases, one hemispheric hippocampus may be surgically removed to treat severe epilepsy or other neurological disorders. The remaining part can be considered a "preserved hippocampus" or a "compromised" one? And traumatized hippocampus from brain injury or stroke? The cases that come from neurodegenerative diseases like Alzheimer's can cause progressive damage to the hippocampus. When is it enough to consider it "compromised"? There is no generally accepted operational definition for a "compromised hippocampus."

Nevertheless, it is not so difficult to define it operationally in terms of the function that is being tested. The concept of a "compromised hippocampus" does not need to be as fuzzy as it may seem at first sight. It may be empirically settled by assessing the severity of hippocampal damage through empirically measured factors (Zeng et al., 2021) that include (i) volume loss (the degree of its shrinkage or atrophy), (ii) functional impairment (the extent to which the hippocampus is unable to perform its normal functions, such as memory consolidation); (iii) disconnection (the degree to which the hippocampus was disconnected given some damage to other brain regions: "severed").

Through these variables, we can establish clear thresholds beyond which no new long-term memory is possible. For example, autopsies or imaging techniques like MRI (Kivipelto, Solomon, & Mangialasche, 2021) can be used to assess the integrity and connectivity of the hippocampus in individuals who are unable to form new long-term memories. Two limits can be established here: the higher the percentage of damaged tissue from which someone can form a new memory, the closer it is to a threshold of a necessary condition. Similarly, the higher percentage of disconnection of neural pathways from which someone could form a new memory would be another threshold of a necessary condition to it, forming, in fact, two different PLs to memory consolidation.

Why should we not consider this condition a contributory condition composing a combinatorial law? We can, but until this threshold (the more damage there is, the more causal pressure against new memory consolidations). Nevertheless, a necessary condition that presents some range of continuous values (from zero damage to the threshold, for example) does not work like a contributory condition. A person can form a new long-term memory without the contributory condition of emotion (only through repetition), but she or he cannot form a new long-term memory without a hippocampus. Therefore, despite the variable of the integrity of a hippocampus working, when within a range of values, as a contributory condition, it is intrinsically a necessary condition to memory consolidation (unlike emotion or repetition).

Beyond the threshold, the integrity of the hippocampus works as a discrete necessary condition. Thinking causally, necessary conditions, especially when perfectly discrete, have a passive character instead of the active one typical of contributory conditions. For example, the more emotion a human experiences in a situation (or repetition of it), the more detailed and stronger the long-term memory formed. However, a person cannot have a "more preserved hippocampus" in a given situation; it is a structural condition, not a dynamic one.[7]

Let us revisit our development psychology example (Hoff, 2006; Kuhl, 2010). We have extracted two PLs from it. First, "if no exposition to language till five, no language at five," which we can represent by $\forall x\ (\neg Ex \rightarrow \neg\Diamond Lx)$, and the second PL is "if no human baby, no language at five" (no talking rabbits). Let us work on the first that is more significant.

No children will acquire language between the ages of two and five if they are not exposed to language on a daily basis (E). It is easy to understand that there will be no talking Kaspar Hauser,[8] no grown human with language having lived in complete isolation. However, what will be the exact threshold? It is evident that insufficient linguistic contact will have no impact on proper linguistic competencies; however, establishing it would be too difficult due to ethical reasons. Therefore, we may opt here for a discrete necessary condition: the exposure to language whatsoever until the age of five. However, such a law would be less informative.

6.2.2.3 *Examples of Possibilist Laws in Psychology*

Candidates to PL, in psychology, are everywhere. We will see here some new examples in psychotherapy, personality psychology, cognitive psychology, and developmental psychology.

6.2.2.3.1 Developmental Law of Abstract Skills

Although it has been updated by decades of experimental research in developmental psychology, Piaget's work remains a significant early achievement in scientific psychology. It set the theoretical and methodological basis of developmental psychology, and among its central principles is an unequivocal PL that says that mastery of logically fundamental cognitive structures is necessary for mastering more complex structures. One of the derivations of this umbrella law is that without mastery of more fundamental concepts of volume and weight, it is not possible to master the relatively most complex concept of density. We can formalize the general law as follows:

$$\forall x(\neg Fx \rightarrow \neg \Diamond Cx)$$

Where:
Fx: mastery of more Fundamental cognitive concepts
Cx: mastery of more Complex cognitive structures
We can read it as "for all individuals x, if x does not master fundamental cognitive concepts, then it is not possible that x will master more complex cognitive structures." The concrete example of it we can formalize as:

$$\forall x(\neg(Vx \wedge Wx) \rightarrow \neg \Diamond Dx)$$

Where:
Vx: mastery of the concept of Volume for individual x
Wx: mastery of the concept of Weight for individual x
Dx: mastery of the concept of Density for individual x
This can be read as: "for all individuals x, if x does not master both the concepts of volume and weight, then it is not possible that x will master the concept of density."

We can be certain that mastery of fundamental cognitive concepts is a necessary condition for mastering more complex cognitive structures by observing the sequential nature of cognitive development as outlined by Piaget. The dependency of complex concepts on fundamental ones ensures that an individual cannot grasp the concept of density without a solid understanding of basics like volume and weight. This has evident practical applications in education, where curricula can be designed to obey the necessary steps on the construction of knowledge, ensuring that foundational concepts are well-understood before moving on to more complex ones. For example, in science education, students should understand volume and weight thoroughly before introducing the concept of density.

Contemporary developmental psychology acknowledges Piaget's contributions as pioneering, but has expanded and refined his hypothesis. It did it through new research methods, interdisciplinary approaches, and a deeper understanding of cognitive development (Case, 1998). Although certain aspects of Piaget's theory have been revised, its fundamental principle remains intact. And it is that foundational concepts and cognitive structures support learning more complex concepts and cognitive structures. Nonetheless, contemporary research has demonstrated that children's understanding of concepts such as volume and weight is influenced by innate cognitive abilities (Spelke, 2000). This shows that human beings possess more innate abilities than Piaget initially suggested. However, we should note that the dependence of understanding density on the concepts of volume and weight has not been challenged.

6.2.2.3.2 *Progressive Exposure*

Without progressive exposure to the object of phobia, an individual can't overcome this specific phobia. We can formalize this as:

$$\forall x\ (\neg Ex \rightarrow \neg\Diamond Ox)$$

Where:
Ex: progressive exposure to the phobic stimulus
Ox: Overcoming phobia of the stimulus in E

We can read this as "for all individuals x, if x does not undergo progressive exposure to a phobic stimulus, then it is not possible that x will overcome the phobia of this stimulus."

This law of psychology is very important and practical. It makes exposure therapy mandatory in cases of phobia and disqualifies other psychotherapeutic approaches to the problem. This law represents a central principle of behavioral therapy: progressive exposure to phobic situations and objects to weaken fear (classical conditioning) associations. It shows that exposure therapy is crucial for treating phobias (Craske et al., 2014).

6.2.2.3.3 *Perception Law of the Size-Distance Relation*

It is the principle that without a correct perception of an object's distance from the observer, it is not possible to correctly estimate its size. Formalization:

$$\forall x(\neg Dx \rightarrow \neg\Diamond Sx)$$

Where:
Dx: correct perception of the object's Distance
Sx: correct perception of the object's Size

We can read this as: "for all individuals x, if x does not correctly perceive the distance of an object, then it is not possible that x will correctly perceive the object's size."

This formulation shows a PL. Depth perception is a necessary condition for size perception. Experimental research in perception demonstrates this relationship, showing that inaccuracies in judging distance lead to corresponding errors in judging size. This interdependency has been well-documented for some time (Rock, 1984).

This relation has practical implications in fields such as virtual reality and robotics, where accurate distance and size perception are critical. An example is programming in virtual reality, where it is always ensured that distance cues are accurately represented to improve the user's ability to perceive the size and distance of virtual objects correctly. Similarly, in robotics, programming robots to perceive distances accurately enhances their ability to interact with objects of various sizes in their environment.

6.2.2.3.4 *Phonological Loop and Word Learning*

The psychology of memory tells us that a minimum limit of functioning of the phonological loop is necessary to learn new words.

Formally:

$$\forall x(\neg Px \rightarrow \neg \Diamond Wx)$$

Where:
Px: sufficient functioning of the Phonological loop
Wx: learning new Words
We can read it as "for all individuals x, if x does not have regular functioning of the phonological loop, then it is not possible that x will learn new words."

The phonological loop is a part of Baddeley's model of working memory. It is critical for temporarily storing and manipulating verbal information. Without the regular functioning of this loop, the ability to learn and retain new words is significantly impaired. The challenge here is empirically establishing the threshold below which no one can learn a new word. On the limit, however, there is no word learning without any phonological loop. This relationship is well-supported by extensive research in cognitive psychology (Baddeley, Gathercole, & Papagno, 1998).

This relation has practical implications in educational and therapeutic settings. Interventions to improve phonological loop functioning benefit individuals with language learning difficulties (such as dyslexia or working memory deficits). Phonological training exercises and memory games are techniques that can improve the phonological loop's performance and facilitate word learning and overall language performance.

Research on cognitive load has shown that overwhelming the phonological high impair its efficiency, affecting word learning (Gathercole et al., 2004). Developmental studies have highlighted that the phonological loop's capacity of children increases with age, correlating with improvements in language acquisition (Alloway, 2009). Furthermore, individual differences in phonological loop capacity have been linked to language proficiency (Conway, Cowan, & Bunting, 2001), and impairments in this loop are associated with language disorders, such as dyslexia (Swanson, Zheng, & Jerman, 2009). These findings collectively support the necessity of the phonological loop for new word learning, reaffirming the PL without falsification hitherto.

6.2.2.3.5 Hierarchy of Needs

This iconic "pyramidal" theory from the psychology of motivation posits that satisfying more fundamental needs (physiological and safety) is necessary for higher-level needs (such as self-actualization) to become motivating. It can be logically represented as:

$$\forall x \forall y(Bxy \rightarrow (\neg Sx \rightarrow \neg \Diamond My)$$

Where:
Bxy: x is a more Basic need than y
Sx: Satisfying need x
My: need y become the focus of Motivation

We can read, "For all needs x and y, if x is a more basic need than y, then if x is not satisfied, it is not possible that y becomes motivating."

The original theory of Abraham Maslow (1954) was a complex PL from the beginning. This posits the foundational role of satisfying basic needs before higher-level motivations can emerge as the focus of our behavior. Maslow believed that more fundamental needs, such as physiological requirements (hunger, thirst, cold, sleep, and breath), must be satisfied before individuals can feel motivated by higher-level needs like self-actualization. His hierarchical structure of motivation is a hierarchy of necessary conditions, setting that without fulfilling basic needs, higher-level aspirations remain silenced. As satisfying basic needs does not guarantee that higher needs will become motivating, the proper representation of this relation is the operator of possibility.

However, even Maslow recognized that some people might prioritize higher needs even before all basic needs are fully met. For example, during a hunger strike, an individual might prioritize a need for justice over physiological needs. Perhaps this is the case because we are not addressing genuine needs, but rather responding to causal pressures or "demands."

Subsequent research has also questioned the rigid sequential interpretation of Maslow's hierarchy, indicating a more complex interplay between different levels of needs (Wahba & Bridwell, 1976). Maslow's original formulation of the hierarchy of needs treats needs satisfaction as binary (either satisfied or not). In reality, needs might be satisfied to varying degrees. And when some basic need becomes increasingly satisfied, the superior need becomes increasingly stronger, in a complex possibilist hierarchy of ceteris paribus laws. However, recent empirical research has also provided support to Maslow's hierarchy. Neurobiological research has shown that brain regions associated with basic survival needs must be satisfied before higher-order cognitive processes related to self-fulfillment can be activated (Kenrick et al., 2010).

Maslow's hierarchy of necessities is, to this day, very influential in fields such as education and management. For example, a manager is educated to ensure that employees' basic needs are met in organizational settings, thereby improving motivation and productivity. Similarly, universities look for satisfying the physiological and safety needs of students in educational environments is fundamental for better learning and motivation to achieve higher academicals.

6.2.2.3.6 Counterfactual Thinking to Regret

The psychology of thinking says that engaging in counterfactual thinking is necessary for experiencing regret in decision-making. We can formalize it as:

$$\forall x(\neg Cx \rightarrow \neg \Diamond Rx)$$

Where:
C(x): engaging in Counterfactual thinking
R(x): experiencing Regret

In simple English, we have "for all individuals x if x does not engage in counterfactual thinking, then it is not possible that x will experience regret."

This is another possibilist law stemming from the general dependence that more complex cognitive operations have on the more basic conceptual components. Considering what might have been is an integral part of the emotional experience of regret. It is through this process that we imagine how better results could have been achieved had we made different decisions (Roese & Olson, 1995).

6.2.2.3.7 Attention to Memory Formation

The psychology of memory says that without attention and encoding, it is not possible to form explicit long-term memories. We can formalize this as:

$$\forall x(\neg(Ax \wedge Ex) \rightarrow \neg\Diamond Mx)$$

Where:
x: information
Ax: x receives Attention
Ex: x is Encoded
Mx: x forms explicit long-term Memory

We can read, "For all information x, if it is not attended to and encoded, it is not possible for it to form explicit long-term memories."

This PL specifies two necessary conditions—attention and encoding—for the possibility of a particular outcome: long-term memory formation. Attention is the cognitive process of selectively concentrating on one set of information while ignoring other perceivable information. Encoding refers to converting sensory input into a form that can be stored and later retrieved from memory (visualizations, verbal descriptions, symbols, diagrams, formulas). Research consistently demonstrates that attention and encoding are necessary to form long-term memories.

Craik and Lockhart's (1972) levels of processing framework indicates that the depth of processing (which includes attention and encoding) directly affects memory retention. They indicate that the information must be deeply processed in order to be transferred into long-term memory. Baddeley and Hitch's (1974) working memory model also shows the role of attention in retaining and manipulating information for encoding it into long-term memory.

Moreover, Paller and Wagner's (2002) studies indicate neural correlates of attention and encoding processes, showing that these cognitive functions activate specific brain regions responsible for long-term memory formation. Their research highlights that the neural mechanisms required for memory consolidation are not activated without sufficient attention and encoding.

Nevertheless, we should always be cautious in psychology (and every science). There are some caveats here. I explicitly use "explicit memory" in the law because some forms of implicit memory, such as the priming effect, can be formed without conscious attention. Moreover, a high level of emotion can automatically encode information into explicit long-term memory without deliberate encoding (but not without encoding).

This PL has several practical applications in education (enhancing students" learning by engaging in encoding), memory improvement techniques, and psychotherapy (helping individuals with memory impairments).

6.2.2.3.8 Universal Grammar to Language Acquisition

According to the psychology of language, human beings possess an innate, universal grammar, which is necessary for language acquisition. Chomsky's (1995) linguistic theory, particularly his recent formulations under the Minimalist Program, focuses on the innate structures and principles that govern language acquisition and use. The Minimalist Program (MP) explain how language works by showing that its basic rules are simple and efficient, shaped by human biology (Chomsky, 1995). A key part of Chomsky's theory is the Universal Grammar (UG). It is the idea that all humans are born with the same built-in language rules, which apply to every language. UG provides the foundational structure that allows humans to acquire any natural language. We can formalize this PL as:

$$\forall x\ (\neg UGx \rightarrow \neg\Diamond Lx)$$

Where:
x: any animal
UGx: x possesses innate Universal Grammar
Lx: x acquires human Language
In plain English:
"For any animal x, if x does not have Universal Grammar (UG), then it is not possible for x to acquire a human language (L).

This PL posits that the presence of Universal Grammar is a necessary condition for the acquisition of any natural human language. It has suffered falsification trials. The most famous is the chimpanzee jokingly named "Nim Chimpsky" (Terrace et al., 1979). Chimpsky was the subject of a famous study in the 1970s aimed at understanding language acquisition in non-human primates. The hypothesis was that primates could not learn language due to the lack of verbal organs capable of producing human phonemes and the absence of linguistic stimulation from birth. To overcome these two obstacles, Chimpsky was raised in a home environment similar to that of a human child, with some researchers and their families, and was taught American Sign Language (ASL) by his caregivers. The leading researcher behind this study was Herbert S. Terrace.

Chimpsky did acquire a substantial vocabulary in ASL, learning to use many signs to communicate his needs and desires. However, his language acquisition was limited to the lexical level. This means that he could use individual words (or signs) to communicate meaning but was not capable of progress in the syntactic level, where grammatical rules and sentence structures are applied to form infinite sentences and communicate complex meaning. His communication remained at the level of simple, context-driven requests rather than complex, grammatically structured sentences. One more time, Chomsky's theory was corroborated.

I should stop here to avoid bothering the reader too much. However, dozens of candidates for PLs occur to me without further investigation. For example, in personality psychology, it can be said that coherence between the assumed narrative and the history of one's life is necessary for effective leadership (Gardner, 1995). In motivational psychology, we can say that satisfying the needs for autonomy, competence, and relatedness is necessary for intrinsic motivation (Ryan & Deci, 2000). In the psychology of thinking, we can say that matching one's skills to the challenge level is necessary for achieving a flow state (Csikszentmihalyi, 1990). In learning psychology, we can say that without observing a model demonstrating a behavior, it is impossible to learn that behavior through social learning (Bandura, 1977). In sum, virtually all psychological scientific production could be partially explicated as PLs.

Note that none of the PL candidates presented are trivial and that their predictions are all useful. Knowing density and compression is insufficient to fully understand volume, so we need to teach them first to make learning volume possible. Coherence between an assumed narrative and your life story will not make you a Moses, but the lack of it will prevent you from being a leader. A severed hippocampus will restrict someone's freedom and memory, but it will not determine it: what they want to remember from their childhood before the injury is up to them.

6.2.2.4 Examples of Possibilist Laws in Neuroscience

PL candidates in Neuroscience are also everywhere. Remember that each link of a causal-mechanistic explanation (that explains capacities and not individual behaviors), if not redundant, implements some PL for that capacity. Let us now see some different examples of these laws in neuroscience:

6.2.2.4.1 Neurotrophic Factors for Synaptic Plasticity

For neurons to perform synaptic plasticity (which is necessary for learning and memory, they must have sufficient levels of BDNF (Brain-Derived Neurotrophic Factor). We can formalize this as:

$$\forall x\,(\neg Bx \rightarrow \neg \Diamond Px)$$

Where:

Bx: Sufficient level of BDNF in neuron x

Px: Synaptic plasticity in neuron x

We can read this as "for any neuron x, if x does not have a sufficient level of BDNF, then it is not possible that x will exhibit synaptic plasticity." This formulation captures a fundamental biological principle supported by extensive neurobiological research. Neurotrophic factors like BDNF are essential (Lu, 2003) for the maintenance and modulation of synaptic connections. Therefore, they are necessary to synaptic plasticity, learning, and memory processes. Without a given level of these factors, neurons cannot proceed with the synaptic modifications required for plasticity. This law helps in understanding various neurological and psychiatric conditions where synaptic plasticity is impaired, such as depression, Alzheimer's disease, and schizophrenia. It also provides a basis for therapeutic interventions to increase neurotrophic factor levels to make neurological change possible.

Despite being a continuous necessary condition, we can easily set a threshold below which there is no plasticity. The BDNF level is not a mere contributory condition; its absence forbids neuroplasticity, and at the limit, it is a radical PL.

6.2.2.4.2 GABA Levels and Anxiety Disorder

Balanced levels of the g-aminobutyric acid (GABA) neurotransmitter are necessary for the regulation of anxiety. We can formalize this as:

$$\forall x\ (\neg Gx \rightarrow \neg \Diamond Rx)$$

Where:
Bx: Balanced levels of GABA
Rx: Regulation of anxiety

We can read this as "for all individuals x, if x does not have balanced GABA levels, then it is not possible that x will regulate anxiety." This PL describes the biochemical basis of anxiety regulation and is supported by research in psychopharmacology.

Evidence supporting the role of a dysfunctional GABA system (Nemeroff, 2003) has resulted from clinical experience with benzodiazepines, as well as neuroimaging studies of the GABA receptor. The results suggest that a deficiency in GABA neurotransmission is always present in anxiety disorders and that all kinds of psycho chemicals that emulate or block the reuptake of GABA possess anxiolytic properties. Despite the fact that the exact character of the influence of GABA in anxiety is disputed (contributory or necessary condition), and the "mechanisms underlying the pathological characteristics of the various anxiety disorders have yet to be fully elucidated" (Nemeroff, 2003, p. 133), the findings are instrumental in pointing substances that can recreate the conditions for patients with anxiety disorders to regain control over these symptoms.

6.2.2.4.3 Dopamine and Fine Control

A minimum level of dopamine in the basal ganglia is known to be a necessary condition to fine motor control. We can formalize this as:

$$\forall x\ (\neg Dx \rightarrow \neg \Diamond Fx)$$

Where:
Dx: minimum level of Dopamine in basal ganglia

Fx: Fine motor control

We can read that as "for all individuals x, if x does not have a minimum level of dopamine in the basal ganglia, then it is not possible that x will have fine motor control."

This is a strong candidate for a PL, as it highlights a well-documented biological necessity in neuroscience. The role of dopamine (5-hydroxytryptamine) in motor control has been well-established by experiments with mice since the fifties (Carlsson et al., 1958), making this a robust and measurable condition. Minimum dopamine levels are necessary but not sufficient alone; other factors like neural circuitry integrity and receptor sensitivity must be satisfied for fine motor control.

6.2.2.4.4 Myelination and Conduction Velocity

Without myelination of an axon, achieving high conduction velocity of nerve impulses is impossible. We can formalize as:

$$\forall x(\neg Mx \rightarrow \neg \Diamond Vx)$$

Where:

Mx: axon x is Myelinated

Vx: Axon x has high conduction Velocity

We can read that "for all axons if they are not myelinated, it is not possible for them to have high conduction velocity."

Since the beginning of the eighties (Waxman, 1980) we have known that myelination is essential for the efficient transmission of electrical impulses through the nerve fibers. It is the process of coating axons with a fatty substance called myelin. Its presence allows electric conduction, where the nerve impulse jumps between nodes of Ranvier, significantly increasing conduction speed. Without myelination, nerve impulses would propagate much more slowly, if at all, thus demonstrating the necessity of myelination for high conduction velocity. New research (Hartline & Colman, 2007) has consistently corroborated this, indicating a relation of necessity likely instantiated by the myelin cover. It is easy to imagine that the kind of isolation that is caused by myelin could be caused by another substance or mechanism, but for practical reasons and based on available evidence, we can consider myelin the necessary condition until proven otherwise.

6.2.2.4.5 Co-activation and Synaptic Connection

It is important for us to consider also falsified PLs to assert their empirical content and scientificity. A traditional implicit PL of neuroscience said that without repeated co-activation of neurons, it is impossible to strengthen the synaptic connection between them (long-term potentiation). We can formalize this as:

$$\forall x\,(\neg Cx \rightarrow \neg \Diamond Sx)$$

Where:

Cx: the group of neurons x is repeatedly Co-activated

Sx: Synaptic connection between neurons x Strengthens

We can read, "for all synaptic connections, if the neurons are not repeatedly co-activated, it is not possible for the synaptic connection to strengthen." Since the now-classic paper by Bliss and Lømo (1973) described the phenomenon of long-term potentiation (LTP), cognitive neuroscience has generally considered that repeated co-activation of neurons is a necessary condition for strengthening synapses between them. Their research showed how high-frequency stimulation of the perforant[9] path in the hippocampus resulted in long-lasting increases in synaptic strength, indicating that repeated co-activation is crucial for LTP. It revealed at least one group of synapses in the hippocampus whose efficiency is influenced by activity that occurred several hours earlier (Bliss & Lømo, 1973, p. 355).

However, recent experiments suggest that other conditions can induce LTP even without typical high-frequency stimulation (Cohen & Greenberg, 2008). Two examples are increased intracellular calcium or pharmaceutical agents that enhance NMDA receptor function. That suggests that while co-activation is a common pathway, manipulating the downstream effects can also induce LTP.

This example illustrates how a hypothesis about a necessary condition can be falsified by demonstrating that what was thought necessary is actually permissive. This opens the way to identify more fundamental necessary conditions that underlie multiple pathways to the same outcome.

Again, none of the PL candidates above are trivial or fail to make useful predictions. Despite this, I must acknowledge that current neuroscientific practice does not seek laws within its overall framework (Craver, 2007). However, this may be partly a consequence of neuroscientists" lack of awareness of the nature of PLs, permissive conditions, and even possibilist thinking. They may become aware of how probabilistic laws can represent, in a simple and falsifiable manner, the necessary relations within the mechanistic models that are their primary focus. However, if this is the case, it depends on whether the possibilist model proves itself useful for neuroscientific inquiry as well.

6.3 Psychological Explanations and Possibilism

The two main traditions of scientific explanation, the causal-mechanical model and the deductive-nomological model (DNM), assume the belief that a phenomenon is only perfectly explained if we offer an answer to the reason why it happens sufficiently, explaining all the causes or laws and conditions that were sufficient for its instantiation. Explanations that are not capable of accomplishing this by virtue of their logical structure—such as how-possibly explanations, covering laws based on ceteris paribus laws, or statistical laws—would, in this case, be only sketches of complete and unrealizable scientific explanations (or just yet unknown).

Two types of phenomena can be the target of explanations: particular events and derived laws. However, what ultimately tests the success of a science is its ability to predict particular events, such as in the case of psychology, individual human behavior.

This can only be done by laws. Explanations are demanded to fulfill some psychological need for comprehension, offering the reasons for something that has happened. They have a task that is easier to do speculatively but perhaps even harder to do precisely since the event can no longer be observed as it happens. Moreover, in psychology, we have no access to any perception of the internal processes of the person who behaved.

In the same way that with laws, the primary object of demand for psychological explanations is individual behaviors, not the capacity of mechanisms or derived laws. This is, therefore, the primary criterion of the adequacy of models in the discipline. Let us evaluate its most influential models under this criterion.

We can divide the different kinds of explanations in psychology by two criteria of classification: their form and their content. The forms of explanations in psychology should be the same as general science; they are the DNM, the probabilistic, the mechanistic, and the possibilist. The content of these explanations is what is unique in the discipline; in other words, they are about what kind of reasons (causes) should be considered when we give a psychological explanation. This second criterion is not our central concern here; therefore, it will only be briefly explained to facilitate the comprehension of the exposition.

6.3.1 Different Explanation Contents in Psychology

Scientific psychology often explains behaviors in terms of Belief-Desire-Intention (BDI), information processing, neuronal activity, or evolution (Cummings, 2010). In other words, when presented, the psychological explanation of particular behaviors generally makes use of implicit laws or some correlate of them as generalizations. They do it for these various levels that, as a rule, are not incompatible but complementary. The option for different contents to explain a behavior or a capacity is a metaphysical option. In choosing one of these levels to explain an event, someone chooses what she or he thinks is fundamental to the phenomenon.

6.3.1.1 Belief-Desire-Intention Explanations

When we explain a behavior based on these three mental states, we call this explanation a BDI. This kind of explanation explains any action with a descriptive combination of what individuals believe, desire, and intend. This framework is deeply rooted in folk psychology, the common-sense approach to explaining actions. It provides a straightforward and intuitive way to understand why people behave the way they do.

Let us use the same act to clearly show the differences in content between these kinds of psychological explanations. Peter runs. A perfectly acceptable BDI explanation of Peter going to run can be done by saying that Peter went to run because he believes that regular exercise is essential for maintaining good health (belief), and he wants to stay fit and healthy (desire), so he decides to go for a run today (intention).

BDI explanations are usually criticized for being supposedly "naïve," as they overlook the complexity of underlying neural mechanisms and environmental influences. Moreover, they generally rely on introspective data, which renders them unscientific. That is not only due to the lack of objectivity, but also because individuals may not always be aware of their true motivations or may not report them honestly. Nevertheless, BDI explanations remain the primary way laypeople and psychologists explain behaviors.

6.3.1.2 Information Processing Explanations

IPEs describe the mind as an information processor, analyzing how information is received, processed, stored, and retrieved. It is a computationalist product, that is, a product from the view of the human mind as a biological computer, processing information through these stages. This approach proved fruitful for understanding cognitive functions such as perception, memory, decision-making, and problem-solving. Let us compare this kind of explanation with the others through our previous example, Peter runs. An information processing explanation would say that Peter considers running if he feels no aches or pains (perception), recalls how good he felt after previous runs (memory), and thinks about his fitness goals (process). So, he decides to run because it will help him meet her fitness goals (decision-making) and plan his run for the morning when the weather is cooler (action).

Human beings are definitely information processors. However, the processing model is sometimes criticized for being too mechanistic. It tends to treat the mind as a machine, supposedly ignoring emotional, social, and contextual factors. Additionally, human thought cannot be reduced to an information process. Therefore, it is more conscious to integrate this kind of explanation with others that capture the other causes of behavior.

6.3.1.3 Neurological Explanations

This kind of explanation presents the brain's functions and structures as causes of behavior and mental processes. This approach focuses on how specific neural processes underpin cognitive functions. Using the same behavior again to clarify the differences, when Peter thinks about running (due to some previous neurophysiological process), his prefrontal cortex becomes activated, helping him weigh the pros and cons of running (involving decision-making). The anticipation of the runner's high releases dopamine, motivating him to put on his running shoes (neurotransmitter release). Neural pathways strengthened by previous runs facilitate his preparation and execution of the run (habit formation).

This kind of explanation faces heavy criticism, not only for being reductionist, but especially for having no relevance to us in our daily lives. Neuronal explanations generally overlook the higher-level cognitive and environmental influences that also play crucial roles in shaping behavior, leaving no room for a conception of consciousness with causal powers. Additionally, interpreting complex neural data and directly linking it to specific behaviors is not possible, at least at this present stage of our knowledge. Despite these issues, neuronal activity explanations are essential for developing interventions in mental health and neurological disorders.

6.3.1.4 Evolutionary Explanations

Evolutionary explanations always explain how certain traits and behaviors that are present now have been selected for their adaptive value. It could have explained Peter's decision to go for a run because, from an evolutionary perspective, regular physical activity would have improved the survival and reproductive success of his ancestors; therefore, he feels an instinct to run. Being fit and healthy enhances his ability to perform daily tasks and maintain good health, which are ultimately advantageous traits selected through evolution for reproduction and the care of offspring.

There are several problems with this kind of explanation. They are always speculative and impossible to test empirically, as they are based on supposed repeated events that occurred thousands and even millions of years ago. Furthermore, these explanations overgeneralize by applying broad evolutionary principles to behaviors that are also shaped by other factors (such as cultural, social, and individual-specific). Above all, these explanations are only possible explanations of our innate propensities as a species and do not provide much insight into individual behaviors.

6.3.1.5 Socio-ambiental Explanations

In addition to Cummings' (2010) four kinds of content for psychological explanations, other approaches have been used throughout history to explain psychological phenomena. However, many of these are either non-psychological or non-scientific.

The most important of them are behavioral explanations, originating from Watson's behaviorist revolution and further developed by Skinner. They are limited to the role of the environment in eliciting observable behaviors through learning processes such as classical and operant conditioning. Although this approach often disregards internal mental or cerebral states, it remains an essential component of a comprehensive explanation of human behavior. It could not be different, as behavior is the source of data from which we can infer mental processes scientifically.

Other, more overtly sociological approaches suggest that the social context in which individuals operate is the ultimate source of causation for human behavior. It examines how social norms, values, institutions, and relationships influence individual behavior. The classic example of a sociological explanation is Durkheim's (1951) work on suicide, published in 1897, which highlights the social determinants of psychological phenomena.

Another influential and relatively new content kind of explanation in psychology is environmental explanation. Environmental explanations point out the impacts of natural and built environments on human behavior. It can examine, for example, how factors such as noise, pollution, access to green spaces, and urban design can contribute to the behavior outcome.[10]

An integrated approach, such as the one proposed by Bronfenbrenner (1979), combines these diverse perspectives. This "socio-environmental" approach encompasses all environmental and social variables that exert causal pressure on human behavior. It provides a holistic understanding of psychological phenomena as a complex interplay of individual learning experiences, social influences, and environmental factors. While these variables can be translated into individual-level variables—as BDI, informational, or neurophysiological ones – we could consider them in an integrative approach to psychological explanation.

6.3.1.6 Other Psychological Explanations

There are other influential models of explanation in psychology, such as humanistic and psychodynamic. Psychodynamic explanations, in the style of Freud, explore unconscious influences, such as the role of defense mechanisms and childhood traumas, in shaping present behavior. In turn, humanistic explanations focus on creativity and free decisions, the search for self-actualization, meaning, and happiness. Despite their contributions, these models are often criticized for their lack of scientific basis, as they do not (and sometimes do not even try to) rely on empirically testable facts, laws, or mechanisms. In the integrative approach proposed at the end of this chapter, however, the roles of humanistic and psychodynamic explanations within a comprehensive psychological framework will be further clarified.

6.3.1.7 Explanatory Pluralism

Each of these explanatory frameworks has its strengths and limitations. Some, more limitations than strengths. However, they highlight different aspects of human behavior and cognition that are all simultaneous in their causation.

That is why we have a historical demand from psychological science that we can call explanatory pluralism, as Dale, Dietrich, and Chemero (2009) do. This highlights the need to consider explanations at multiple levels for behavior and cognitive processes, as multiple causal factors from different levels of organization invariably influence these. These levels can even exhibit emergent properties (such as consciousness), which can be analyzed in their own terms, without recourse to the more fundamental processes from which they emerged (complex molecular organization and brain cellular activity), as they generate measurable properties of crucial epistemic value.

BDI explanations can be formally specified in terms of information processing, which in turn must be implemented by a physical system (in this case, the brain). Evolutionary explanations can account for why some innate cognitive processing mechanisms or motivational dispositions function as they do, but they offer little insight into individual and unrepeatable behaviors. When an evolutionary belief is used to explain individual behavior, it becomes the reason for some desires and intentions of a BDI being as they are.

My conclusion is that integrating these contents should be an aim of psychology. Such kind of integrative approach would be able to provide a more comprehensive explanation of psychological phenomena. However, this is extremely difficult. How can we do it without becoming an unscientific, naïve eclectic? I will submit my proposal to address this problem in the subitem regarding multicausality and possibilism.

6.3.2 Different Explanation Forms in Psychology

The most important considerations regarding psychological explanations are those related to the application of common models from modern science in the discipline. Scientific psychology can offer explanations in the form of deductive nomological, probabilistic, mechanistic, and possibilist models. Let us consider some metaphysical implications, epistemic limitations, and strengths of these forms in psychology.

6.3.2.1 Deductive-Nomological Model in Psychology

In a brief exercise of imagination, let us focus here now on establishing what a strict psychological deductive-nomological explanation (DNE) would look like based on the most used explanatory strategies in contemporary psychology.

A DNE in psychology, in principle, would not need to specify particular laws and conditions at all levels of variables. If we had sufficiency laws (strict laws) in psychology that necessarily and sufficiently would imply behaviors in the presence of one or a few antecedent conditions, we would have good DNEs based on even a single law. But in psychology and neuroscience, we do not have a single one of these laws. We do not have any, after more than one hundred and forty years of experimental research in the area. For these and other reasons, Donald Davidson (1980) defended the thesis that it is not possible to have strict psychological or psychophysical laws, as there would be no way of specifying the conditions for a universal proposition to be fulfilled without exceptions.

Thus, by the nature of psychology, a DNE of a particular human behavior (the explanandum) should include, in the explanans, general laws governing cognitive functioning, along with all initial conditions related to them (the input information, current beliefs, current goals, etc.). A clearer idea of what such an explanatory model entails can be seen in the following schema (adapted from Castañon, 2007):

If the universal laws L are:

L1) A set of laws concerning human motivation,
L2) A set of laws governing the acquisition of beliefs,
L3) A set of laws regarding universal cognitive capacities,
L4) A set of laws describing how the brain implements cognitive capacities,

And the contingent facts are:

C1) The current state of an individual's goals,
C2) The current state of the individual's belief set,
C3) The current state of the individual's cognitive structures,
C4) The state of the implemented capabilities and the neurochemical balance in the individual's brain,
C) The set of information present in the environment,

If the universal laws L are:

L1) Set of laws on human motivation,
L2) Set of laws on the acquisition of beliefs,
L3) Set of laws on human cognitive capacities,
L4) Set of laws on the implementation of capabilities by the brain,

and contingent facts:

C1) State of an individual's set of goals,
C2) State of the set of beliefs of an individual,
C3) State of the set of an individual's cognitive structures,
C4) State of the capabilities implemented and the balance of substances in an individual brain,
C) the set of information present in the environment,

---

then behavior B occurs.

Well, this is totally unrealizable. If we only consider explained behavior that could have been predicted, not a single particular behavior would be capable of scientific psychological explanation. It is evident that we are not facing a phenomenon like a chemical reaction. The prediction of individual behavior would be, even from a deterministic perspective, the most complex among those conceivable in the universe (excluding the prediction of the behavior of many individuals interacting in society …).

Cummings (2010) notes that, although there is still considerable "lip service" dedicated to DNM, current theories and explanations in psychology are offered without recourse to laws, being designed in terms of functional analysis. Faced with a sterile scenario like this, psychology is divided between the most diverse explanatory strategies. From the attempt to make finalist explanations compatible with science (Rychlak, 1994) to promises of future DNEs with eliminative materialism. Moving away from the strict DNM, we currently have in psychology the probabilistic explanation and the mechanistic explanation. However, as we are about to ascertain, none can explain the necessity (if this exists) of individual behavior.

One popular strategy among materialists faced with the failure of traditional DNE is an escape to speculative reductionism, preached by theories such as eliminative materialism (Churchland, 1981). For it, the development of neuroscience will inevitably eliminate terms from the vocabulary of belief-desire-intention psychology (folk psychology), replacing them with references to patterns of neural activity. The problem is that translation patterns between brain states and folk psychology terms would have to be established for the former to make any sense in psychological terms to common people. In other words, those kinds of people, like me and probably you, who believe that they raised a glass because they wanted to.

Always appealing to a distant future where all these multiple levels of causality will turn out to be simplistic linguistic illusions, faith in the DNM and determinism is maintained based on the pure promise of materialist truth that will be revealed in the future. This glorious future, as we know, is always in brutal contrast with a poor present without relevant achievements. Bechtel & Wright (2009) show that the call for the future reduction of psychological laws and terms to more basic laws of biology is currently incomprehensible, since we also have very few relationships with the force of laws in physiology and biology. And yet Wright (2007) reminds us that successful reductive explanations would not impede explanations of behaviors based on higher psychological processes.

6.3.2.2 Probabilistic Explanations in Psychology

Hempel (1965) realized from the beginning that not all accepted scientific explanations are deductive-nomological; some are merely probabilistic. This model of explanation, which he named the inductive-statistical model (ISM), is an argument that has among its premises (explanans) at least one general statistical regularity (a statistical law) and an initial condition. The conclusion (explanandum) is established, however, not by a deduction, but by mere probabilistic induction.

However, explaining why an event had a high probability of occurring is not the same as explaining why it occurred. Let us consider an example from Cartwright (1999) of attempting to explain a particular event that is subsumed by a probabilistic law. Suppose we have a probabilistic law that predicts 62 percent of camellias will die and 38 percent will survive when planted in the type of soil I have in my garden. This camellia here, in the left corner of the garden, survived. What explains why this particular camellia survived, rather than perished, if all we know is that the probability of survival was 38 percent?

Probabilistic laws are useful because they predict directions of behavior in population samples, and the ceteris paribus laws themselves can express some probabilistic magnitude. But the probabilistic explanation is useless in the particular case. This is Dilthey's (1945) criticism of this type of explanation: extracting anything from them about the individual behavior of a real person is an induction without logical support. Thus, in its probabilistic version, the DNM has a logical downgrade accompanied by the same aggravated defects.

6.3.2.3 Mechanistic Explanations in Psychology

As we have already seen here, the mechanistic explanation has arrived with force in psychology and neuroscience (Leite, 2021; Bechtel, 2008; Craver, 2007). To it, the explanans in psychology would be an adequate specification of a mechanism and the explanandum the capabilities and effects of it. Thagard (2003) classifies this type of explanation as an answer to a "how-question" and not a "why-question." CM explanations in psychology, therefore, fulfill only one of the two functions proposed by the DNM: they explain derived laws (effects), not individual behaviors.

Authors such as Leite (2021) even interpret mechanistic psychological explanations as a new school, a new metaphysical approach to the discipline. He calls it (p. 39)

"The Mechanistic Theory of Human Cognition" (MTHC). It would be a psychological counterpart to neomechanicism, a physicalist philosophy of psychology based on work in contemporary cognitive science. I do not think that what Leite calls MTHC is a new school or even a new kind of content in psychological explanation. It is the neurological explanation all over again, presented in mechanical form. Let us, then, analyze this form, its strengths and its difficulties.

Fodor (1991b) argues that the central goal of theory building in cognitive science is to explain the computational mechanisms that implement intentional laws[11] (generalizations that relate propositional attitudes like beliefs and desires to actions). Such laws can be represented in the form of algorithms: structured sequences of "if then" rules (function) that, when executed, produce exactly that effect (behavior). For example, a typical intentional law might state: if an agent believes that pressing the button will open the door, and desires to open the door, then the agent will press the button (ceteris paribus). The computational theory aims to specify an internal mechanism that processes such beliefs and desires, producing the appropriate behavior. The problem is to apply this explanation to individual cases: identifying all the specific antecedent conditions that, in interaction with the mechanism, produced a particular behavior in a given person at a given time. Although the many intervening variables (like distraction or conflicting desires) make this task pragmatically difficult, it is not logically problematic if the mechanism were a well-specified function.

In another example of the problem, an algorithm would need to specify how the visual system perceives colors in order to produce the afterimage effect of opposite colors. Although such after-effects do not involve intentions per se, they highlight the same challenge I am addressing here: the difficulty of specifying all the real-life antecedent conditions that, together with the mechanism, yield a particular psychological outcome at the level of an individual case.

The problem is that, by rejecting psychological explanation as the deduction of phenomena through laws, authors such as Fodor (1991a), Bechtel & Wright (2007), and Cummings (2010) do not recognize that a mechanism works as a large and complex law (if it is a function). Moreover, they do not offer an alternative logical model of explanation, limiting themselves to repeating that the mechanism holds a privileged place as the explanans, and the law is merely an effect that constitutes the explanandum. We do not even have a reductive analysis of the concept of mechanism, let alone of mechanistic explanation. Bechtel and Wright (2009) even state that diagrams and figures often represent mechanisms more effectively than linguistic descriptions.

The discipline rejected DNM because it was judged unattainable. However, how attainable is the mechanistic explanation? The task of constructing an ME in psychology demands experimental procedures and manipulation since the complexity of psychological mechanisms, if they exist, does not reveal their functioning to an observer. More than this, they are, by nature, always occult.

In turn, manipulating a mechanism means intervening in its everyday operations to test the functioning of its parts and organization. As Leite (2021) points out:

> There are two main ways of intervening on a mechanism: either by modifying its input or the external conditions under which it functions; or by modifying the internal operations of the mechanism itself. Consequently, a change can be detected at the level of the general function of the whole mechanism or at the level of the functions of its components. In behavioral experiments, made in neuroscience and cognitive neuroscience, for instance, the idea is to manipulate the task posed to the organism or to the subject in order to understand the mechanism that is responsible for that behavior. Thus, this kind of experiments do not offer a direct measure of the internal mechanistic structure itself or of any of its operations. (p. 31)

Therefore, the "hard science" of neuroscience does not usually offer any direct observation or measurement of the supposed mechanisms; it only recurs to psychology to measure the subject's performance in a specific task and its variations, usually by its results, reaction time, or time to complete the task, correlating this to images of the brain.

According to Bechtel (2008, p. 41), the only two forms of internal intervention that can be used are, first, temporarily impairing the component (electromagnetic stimulation) or permanently impairing it (lesion experiments). Second, stimulation of mechanistic components increases their performance or likelihood of occurrence.

Some experiments measure internal brain operations under normal conditions, investigating their location and electrical and chemical activity (PET, fMRI, or MEG), relating these to cognitive operations and sensory stimulations. Again, in these cases, the subjects perform a cognitive task during a PET scan or a magnetoencephalography. Studies with single cells are unusual in humans because they are an invasive and potentially harmful method that requires exposing the brain and destroying the cells being studied.

The problem with these techniques is straightforward. They are not direct measurements of neuronal activity, and neuronal activity is not a direct measurement of psychological activity. We have two levels of distance from the real object. Positron Emission Tomography (PET) measures the signal by detecting gamma rays emitted by a radioactive tracer, typically a form of glucose, that is administered intravenously. The presupposition is that the consumption of glucose is associated with cerebral activity used to fulfill the task. Functional Magnetic Resonance Imaging (fMRI) indirectly measures neural activity through changes in blood flow. The presupposition is that this flux increases due to increased neuronal activity. Finally, a more recent technique called Magnetoencephalography (MEG) measures the magnetic fields associated with electrical activity in neurons.

Nevertheless, this is the easy part. All these techniques have spatial or temporal distortions, meaning they cannot determine exactly where the activity is happening or when. A PET scan has poor temporal and spatial resolution. fMRI has good spatial resolution but moderate temporal resolution. MEG has excellent temporal resolution but moderate spatial resolution. Therefore, even the results about neuronal activity, which cannot be associated with any particular psychological experience, must be analyzed carefully.

This process appears to be quite complex, challenging, and largely out of reach nowadays. Is there an actual example of this in the discipline? Bechtel (2008) offers two examples: visual and memory systems. Let us look at the most developed of them: visual perception.

According to Bechtel (2008), the mechanistic investigation of visual perception begins with identifying the brain regions involved. He provides the example of the occipital lobe, specifically the primary visual cortex (V1 or Broca's area 17), which is central to the vision. Damage to the occipital lobe results in visual processing deficits (not only in humans). The optic tract projects from the eye to the lateral geniculate nucleus (LGN) in the thalamus and then to the occipital lobe. Continuing, Bechtel (2008) shows that visual perception is dependent on the posterior parietal cortex and the inferotemporal cortex, located in the temporal and parietal lobes.

Each brain area involved in visual perception performs different functions. V1, the primary visual cortex, processes basic visual information such as edges, light and dark spots, and bars of light. Other areas, from V2 to V5/MT, encompass process aspects such as color, depth, motion, and the location and shape of objects. Understanding the specific operations of these components involves determining how they process light, color, motion, and other visual features. It is made through electrophysiological recording,[12] lesion studies, neuroimaging (PET, fMRI), stimulation, and psychophysical experiments.

The model also maps the pathways through which visual information is transmitted. This includes the initial reception by photoreceptor cells (rods and cones) in the retina, followed by transmission through the optic nerve to the LGN and then to various cortical areas.

Despite its limitations, the mechanism of visual perception is indeed a strong case for mechanistic explanation in psychology. After all, "Rome wasn't built in a day." Of course, many operations of involved brain areas are still unknown, but as a model of information processing, the progress is undeniable.

However, when we recall that this is the most robust case for a mechanistic explanation in psychology (Bechtel, 2008), despite the immense obstacles to its empirical testing, a question arises: Are mechanistic explanations in a better condition regarding attainability in psychology than DNM?

Unsurprisingly, we do not have one complete mechanistic explanation of any psychological capacity. This does not mean mechanistic explanations in psychology are wrong. This only means psychology is a difficult science. Nevertheless, most of the case against DNM are based on its qualities, clear structure, and evaluation criteria. In mechanicism, nowadays, researchers can attribute psychological processes to the brain without knowing anything about how they are physically implemented and without being held accountable for it.

6.3.2.4 Possibilist Explanations in Psychology

The fourth kind of explanation form in psychology is the possibilist. It is an explanation of how a psychological event was possible. As this item is here only for the sake of logical organization of the exposition, to locate possibilist explanations (PEs) within a different kind of form of psychological explanation, and as it is this section's central objective, I will open a new item only for it..

6.3.3 Psychological Possibilist Explanations

In psychology, PEs are the only ones that truly explain individual behavior, albeit they only explain why it was possible, not necessary. The reason for this is simple. PEs are deductive nomological explanations that work with the only strict law in the universe (PL) and with one condition only to do their job.

6.3.3.1 Explaining Events Just Not Necessary

More important than using PEs for unusual psychological events is their use for common psychological events that are simply not sufficiently explainable. For example, let us discover what could have been done in developmental psychology. How could Piaget (1965) have provided a sufficient explanation for a specific behavior (Event E) that expresses an understanding of the principle of conservation of quantity (the quantity of a thing remains the same regardless of any change in its appearance or shape) in a given situation? This is indeed impossible. The most that Piaget's laws of development could explain is that, due to the presence of a specific cognitive structure, this behavior was possible but not necessary. For example (Castañon, 2021b):

1) For a child, having entered the pre-operational stage of development is a necessary condition to control the principle of conservation of quantity;
2) The mastery of the identity principle necessarily depends on the pre-operational stage;
3) The four-year-old child, Julie, showed efficient use of the identity principle at T1;
4) Therefore, Julie was in the pre-operational stage of development at T1;
5) Therefore, it was not impossible for Julie to present behavior that presupposes control of the principle of conservation of quantity in T2 (event E).

Being in the pre-operational stage is not the only necessary condition for the appearance of the principle of conservation of quantity in a child (the principle of reversibility, for example, would be another). However, in some cases, it may refute the assumption that the child could not have exhibited this behavior then and will necessarily be part of a complete ideal explanation of his behavior.

Regarding using the possibilistic framework in psychological mechanistic explanations, let us look at Fodor's (1991a) example of the cognitive psychology of expertise.[13] Repetition facilitates the information retrieval route, and rehearsal improves performance. We explain these generalizations or "effects" based on the logical structure of the mechanism that produces the ability to memorize facts and develop skills. Nevertheless, of course, that is not the whole story. This is not a sufficient cause for a great pianist to be a great pianist. However, based on what we know about how skill memory works, we can certainly say that spending a lot of time each day playing scales is a necessary condition for being a great pianist. Based on this LP, we can offer a mechanistic, neither deterministic nor voluntaristic[14] explanation: a PE. It would be that, given the mechanism used to develop the skill of playing the piano, one of these stages was completed. If the others, as far as we know, were not abandoned, then the presented behavior was possible.

Let us go back to the example of Piagetian theory. A DNE could never predict behavior that manifests an understanding of the principle of conservation. That is, in fact, impossible due to the complex set of laws and conditions it would demand. The most that Piaget's laws of development can predict is that given the presence of a particular cognitive structure, such behavior is possible, not necessary. Faced with a researcher who asks a question about an operation with volumes involving the principle of conservation, a child can make a mistake in the task, refuse to do it, deceive the researcher by pretending to make a mistake, and so many other possibilities that it is impossible under any circumstances to predict what she will necessarily do. Nevertheless, we can predict a series of things that, under certain circumstances, the child is not capable of doing and many others that it is already possible for him to do; subject these predictions to tests and, therefore, corroborate them.

6.3.3.2 Examples of Possibilist Explanations in Psychology

Having "explained the explanation," let us examine and formalize several examples of PEs in psychology, and see how they fit to deliver what mechanistic and sufficiency DNMs cannot: explanations of individual behaviors.

6.3.3.2.1 The Overcoming of a Phobia

Let us imagine a boy with a pathological fear of spiders. It comes a day that you see him at a school's science fair, presenting a group of spiders on his desk. How can we explain this individual behavior?

An explanation could be that, looking for professional help, he went under cognitive-behavioral therapy. The psychologist had applied a program where the boy was progressively exposed to spiders, first in his imagination, then in images, and finally with real spiders at a distance, and then with a large spider at close range. After this two-week program, the boy has not presented phobic behavior anymore.

Why is this a PE? Because what it does is show a permissive condition for a necessary condition to overcome a phobic behavior, but not a sufficient condition. What it did was eliminate the surprise in the face of the boy's behavior. Let us break down the covert structure of this explanation:

Context (S): Normal cognitive conditions of the boy, normal socioeconomic conditions, and normal medical conditions. A safe and controlled therapeutic environment.

Formal Structure:

$$\forall x\,(\neg Nx \rightarrow \neg\Diamond Ex)$$

(PL: no repetitive exposure, no possibility of lowering fear)
In S, Pa→ Na
(in an S context, P is permissive: behavior therapy)
In S, Pa
(The permissive condition P is satisfied: the boy goes to therapy)
∴ In S, Na
(Necessary condition N is satisfied: non-traumatic exposure to the boy)
∴ In S, ¬(¬◊Ea)
(The lowering of the boy's fear was not epistemically impossible)
Spelled:

In context S, we believe that all necessary conditions N for an individual (a) to overcome a phobia of spiders (E) are satisfied but one: progressive exposure to spiders (Nm). If an individual (x) does not have non-traumatic exposure to spiders, it is not possible for that individual to overcome the phobia (¬◊Ex). Non-traumatic (progressive) exposure to spiders is a necessary condition (N) for overcoming the phobia (E). In context S, the individual being in a safe and controlled therapeutic environment of cognitive-behavior therapy is a permissive condition (P) that allows the individual to undergo progressive non-traumatic exposure to spiders (Na). The individual is in a safe and controlled cognitive-behavior therapy therapeutic environment (Pa). Therefore, the individual undergoes progressive exposure to spiders (Na). Therefore, it was not impossible for the individual to overcome the phobia of spiders (¬(¬◊Ea)).

6.3.3.2.2 The Formation of an Explicit Long-Term Memory

John remembers an episode in his five years when he observed a kissing bug in his uncle's garden. How can we explain this?

He was a healthy boy with no cognitive impairments, and the sight of a big kissing bug impressed him. He was in the garden examining the bug and comparing him mentally with bugs he had seen before, when his aunt came and told him that that bug could carry a disease that could make the heart grow and weaken. That description impressed him even more, making him exam every aspect of the bug. That is why it was not surprising that he remembered this event of his infancy now.

Given:

x: any individual
j: John
E: Encoding of the information
A: Attention to the information.
M: Long-term memory.
We can formalize this explanation as:

$$\forall x\,(\neg Ex \rightarrow \neg\Diamond Mx)$$

(possibilist law)
In S, $Aj \rightarrow Ej$
(in an S context, A is permissive)
In S, Aj
(The permissive condition A is satisfied)
$\therefore$ In S, Ej
(The necessary condition E is satisfied)
$\therefore$ In S, $\neg(\neg\Diamond Mj)$
(The memory was not epistemically impossible)

Where:

In context S, we believe that all necessary conditions for an individual (x) to form a new explicit long-term memory (M) are satisfied, except one: effective encoding of the information (E). If an individual (x) does not encode the information effectively, it is not possible for that individual to form a new explicit long-term memory ($\neg\Diamond Mx$). In context S, the individual paying attention to the information (Ax) is a permissive condition that allows for effective encoding of the information (Ex). John pays attention to the information (Aj) of a bug. Therefore, John encoded the information effectively (Ej). Therefore, it was not impossible for John to form a new explicit long-term memory of the kissing bug ($\neg(\neg\Diamond Mj)$).

It is, indeed, a feeble explanation of how John formed a long-term memory. However, it is a base for constructing complexes and informative explanations. In the subitem about multicausality, we will see a complexification of this explanation with the addition of contributory causes and other necessary conditions.

6.3.3.2.3 An Insight Solving a Problem

I recall an occasion in the technical high school of electronics when my digital electronics professor offered my class a point in the bimestrial grade for solving a problem that involved creating a circuit to execute a problematic function. All my classes became excited about the challenge, and I dedicated several hours with my best friend to solve that problem. Nobody could. The day after, when I was taking a shower, the correct circuit popped into my mind. That was possible because I was in a relaxed and open state of mind at the time, which meant I was not fixated on the wrong approach to solving the problem.

Given:
a: the individual (I, in this case)
Necessary Condition (R): to be in a Relaxed and open state of mind.
Context (C): I had the relevant background knowledge and experience.
Permissive Condition (S): In context C, taking a Shower was relaxing
Explanandum (I): a has an insight of the correct circuit (solution)
We can formalize this PE as follows:

1. $\forall x (\neg Rx \rightarrow \neg \Diamond Ix)$

   (possibilist law)

2. In C, Sa → Ra

   (in context C, S is permissive)

3. In C, Sa

   (The permissive condition S is satisfied)

4. ∴ In C, Ra

   (The necessary condition R is satisfied)

5. ∴ In C, $\neg(\neg \Diamond Ia)$

   (The event I was not epistemically impossible)

6.3.3.2.4 An Attainment of a Flow State

In the process of writing this book, I achieved a flow state several times. How would we explain it possibilistically? There are several ways. I will choose the need to engage in a challenging activity for an individual to achieve a flow state.

Context (S): I was in a good condition of wealth, intrinsically motivated, and sometimes free from distractions.

Permissive Condition (B): being writing the Book.

Necessary Condition (C): Challenging yet achievable activity.

F: Flow state

Formal Structure:

1. $\forall x (\neg Cx \rightarrow \neg \Diamond Fx)$

   (possibilist law)

2. In S, Ba → Ca

   (in an S context, B is permissive)

3. In S, Ba

   (The permissive condition B is satisfied)

4. ∴ In S, Ca

   (The necessary condition C is satisfied)

5. ∴ In S, $\neg(\neg \Diamond Fa)$

   (The event was not epistemically impossible)

In context S, we believe that all necessary conditions N for an individual (a) to achieve a flow state (F) are satisfied, except one: the activity being challenging yet achievable (C). If an individual (x) does not perform an activity that is challenging yet achievable (Cx), it is not possible for that individual to achieve a flow state ($\neg\Diamond Ex$). In context S, writing this book is, to me (a), a permissive condition (Ba) that allows for performing an activity that is challenging yet achievable (Ca). In context S, I wrote this book (Ba). Therefore, I performed a challenging yet achievable activity (Ca). Therefore, achieving a flow state was not impossible ($\neg(\neg\Diamond Fa)$). A caveat here is useful. It is not writing itself that results from the flow state, but rather writing in a particular way—effortlessly and quickly.

Again, we have here a very poor explanation of the possibility of a flow state. Again, we will see that it is a base from which we can offer progressively complex explanations.

6.3.3.2.5 Pirahã Language and Recursion

Daniel Everett's (2005) research on the Pirahã language, spoken by an indigenous group in the Amazon rainforest, suggested initially that this language lacks the feature of recursion. How could this be possible? This absence seemed to challenge the universality of recursion as part of the Universal Grammar as proposed by Chomsky.

However, further research by the same Everett (2012) revealed that, while the Pirahã language does not employ recursion in the same way other languages, it achieves similar communicative functions through alternative structures such as prosody, intonation, and contextual cues.

By identifying these permissive conditions and applying the PL, a PE can show why the Pirahã language does not represent a falsification of the Universal Grammar despite initial interpretation.

Where:
x: any language
a: Pirahã language
Permissive Condition (P): Prosody, intonation and contextual clues.
Necessary Condition (R): Recursion
L: Human Language
Formal Structure:

1. $\forall x\,(\neg Rx \rightarrow \neg\Diamond Lx)$

   (possibilist law)

2. In S, $Pa \rightarrow Ra$

   (in an S context, P is permissive)

3. In S, Pa

   (The permissive condition P is satisfied)

4. ∴ In S, Ra

 (The necessary condition R is satisfied)

5. ∴ In S, $\neg(\neg\Diamond La)$

 (The event was not epistemically impossible)

Breaking down this explanation, we start with a PL ($\forall x\ (\neg Rx \rightarrow \neg\Diamond Lx)$) stating that "For any language x, if it does not have recursion (R), then it is not possible for it to be a human language (L).

In the context (S) of the Pirahã language (a), the use of prosody, intonation, and contextual cues (P) fulfills the role of recursion (R).

Therefore, in the context (S) of the Pirahã language (a), the necessary condition (R) of recursion is functionally satisfied through alternative means.

Therefore, the Pirahã language (a) being a human language (L) through alternative mechanisms was not epistemically impossible.

In other words, using Everett's findings in a PE we can see how the Pirahã language fits perfectly within Chomsky's Universal Grammar framework.

These examples illustrate the application of PEs in psychology, identifying necessary and permissive conditions for various psychological phenomena. They explain individual behaviors, not capacities or laws. Indeed, they do not explain why it sufficiently occurs, but how it was possible is some context, given that it did not happen in others.

These are indeed much more modest claims to scientific explanation. However, in the case of psychology, these are also much more honest. Complete DN or mechanistic explanations of a behavior are practically impossible and metaphysically compromised. On the contrary, the PE of a behavior is possible and metaphysically neutral toward determinism. It is only metaphysically committed to the thesis that the object of psychology is regular in at least some of its aspects, and, as we will see in the next item, it solves the problem of psychological multicausality (cognitive, neuropsychological, genetic, and environmental levels).

6.4 Possibilism's Impact on Scientific Psychology

Having investigated in detail the nature of PLs and CCPLs and PEs in psychology, it is time to return to the obstacles listed at the beginning of the chapter and evaluate the possible impact of possibilism on them.

6.4.1 Possibilism and Traditional Problems of Psychology

Regarding Kant's argument about the unquantifiability of psychology, possibilism offers one more step toward "non-quantitative mathematics" (Baars, 1986) of its laws and explanations. PLs are non-quantitative relations between a necessary condition and some kind of phenomenon. Of course, it offers nothing in terms of the methodological necessity of data measurement. All the difficulties associated with quantifying psychological variables persist.

The problem of explaining or predicting the alteration of the object of psychology by social interaction and knowledge is partially solved by possibilism. They are two slightly different arguments. To Gergen's (1973) argument—that the dissemination of psychological theories modifies the patterns of behavior of those who know them—a PL, which is a genuine strict law, would rule out the possibility of a behavior occurring in the absence of a necessary condition. As such, no behavioral change resulting from knowledge of the theory would be possible unless the necessary condition were already satisfied. That becomes evident in neuropsychological and developmental cases. For example, a person with severely damaged hippocampi cannot form new long-term memories, no matter how much psychological knowledge this person acquires. Similarly, a child who has not yet learned how to add cannot execute an exponentiation properly, no matter how much we expose this child to mathematical theory. Rather than falsifying the law, successful instruction or socialization that enables such abilities would confirm that the necessary conditions had been fulfilled. These examples illustrate that, under possibilist reasoning, social interaction and knowledge can only enable new behaviors if they help satisfy the preconditions for those behaviors—they cannot override the laws that define what is possible in their absence.

About the considerations of Neisser (1975), that the more knowledge we have, the more unpredictable we become, it does not change, but assures the same predictability as pointed out above: if a person with severed hippocampi knows the meaning of this condition previously to the lesion, he can perform some previous known technique to help deal with the impairment, but still, we can predict that he will not be able to form a new explicit long-term memory.

Finally, PLs and explanations offer a radical solution to the major ontological obstacle to scientific psychology. When we predict that knowing addition is a necessary condition for performing a multiplication operation, we are not predicting that a child will perform it now, nor that she will perform it correctly. We are only predicting capacities or incapacities: not determining what the behavior must be, only what it must not be: it must not be the one for which the necessary condition is unfulfilled. A comprehensive scientific psychology based on PLs and causal pressures from CCPLs would provide a scientific description of the limits and conditions of human freedom. In other words, a science of human freedom. We will revisit this supposed solution in depth at Chapter 7.

To those who defend the notion that psychology's object is meaning and that it should employ an idiographic method, possibilism offers a radically new approach, as we will see in the subitem dealing with the discipline's fragmentation. First, however, I will fulfill my promise to, through possibilism, offer a solution to the study of psychology, given its multicausality problem.

6.4.2 Psychology's Multicausality and Possibilism

In the sea of variables that interfere in the supposed determination of behavior, we cannot establish which ones actually determined it and to what extent. Without an adequate solution to this issue, modern psychology would continue to be considered by many as probabilistic fiction without much contact with concrete explanations in the real world.

Multicausality in psychology is far more complicated than several variables of the same kind competing vectorially at the same level of reality to determine, conjointly, the outcome phenomenon. Multicausality in psychology also involves several levels of reality with different variables exerting causal pressures on the same phenomenon. We have at least four: conscious decisions and attributions, sociological, informational, and neurobiological.

For the first one of those four, some believe that no laws can govern it; this would be the source of our relative autonomy and freedom. In contemporary philosophy of mind, an explanatory gap between the set of beliefs and desires and conscious decision is defended by philosophers such as Searle (2001), who considers freedom of volition a necessary condition of rationality. The problem is that if such a conception of the mind is true, the difficulties in constituting scientific psychology are evidently aggravated. The principle of the regularity of the object is at stake here: How would a free and creative human being allow the setting of stable cause-and-effect relationships on it?

As we can see, the problem of multicausality in psychology has always seemed hopeless.

Daniel Robinson (1985) gives a highly suitable example to illustrate my description. Assuming the explanandum, "Adam Smith wrote 'The Wealth of Nations,'" there would be at least nine explanations for what led him to perform this action. He was brought to it because of (1) the physical condition of his brain, (2) unconscious motives, (3) his history of reinforcement, (4) neurotic disorders, (5) having a certain personality, (6) not being distracted by other things; (7) his legitimate motivation; (8) possess sufficient intelligence; (9) having participated in discussions about economics in childhood. Robinson (1985) states that all nine explanations fail individually and collectively "to explain the Wealth of Nations sufficiently" but should figure in a complete explanation of it. However, although they cannot be taken individually as determinative, each condition could be considered "permissive" (Robinson, 1985, p. 61) of the phenomenon.

However, an eventual generalization of the thesis that the laws in psychology are only of two basic types, the possibilist, and the CCPLs, and that all the feasible explanations (DN and mechanistic) in the discipline are possibilist in essence, could offer a valuable tool to solve the scientific problem of representing—limitedly but faithfully—the psychological problem of multicausality.

As we have already seen, an eventual adoption of the PE in psychology would allow the offering of explanations with a simple and precise format, as it would be enough, at the limit, to determine one only variable to explain our explanandum without compromising metaphysically this explanation with a supposedly fundamental nature of this variable.

This happens because the condition it states is only necessary to the phenomenon, not sufficient. In the same way, CCPLs only represent causal pressures; they express contributory, not sufficient conditions.

Because of this, PEs work like the "pegs" Searle (1958) once wanted for proper names. We can always hang new necessary conditions and causal pressures on them, enriching the explanation of why a person has behaved as he did, although we have never determined it sufficiently. This feature allows for PEs to include necessary and contributory conditions from all levels of psychological causation precisely because, for its own nature and the nature of PLs and CCPLs, it is incapable of determining the level from which comes the causal force that, added to the necessary and contributory conditions explained, will sufficiently cause the phenomenon.

Let us start illustrating this characteristic by complexing a previous example a bit. Then, let us return to Fodor's example of musical expertise.

6.4.2.1 Again, the Formation of an Explicit Long-Term Memory

Let us reexamine the same example we examined last time. We should explain how John remembers an episode from five years ago when he observed a kissing bug in his uncle's garden.

He was a healthy boy with no cognitive impairments, and the sight of a big kissing bug impressed (I) him. He was in the garden, examining every aspect of the bug, mentally comparing it with bugs he had seen before, when his aunt came and told him that the bug could carry a disease that could cause the heart to grow and weaken. That description impressed him even more, and since then, every time he thinks of diseases transmitted by bugs, this is the example that pops into his head. That is why it was not surprising that he now remembered this event from his infancy.

Here, we have a combination of the PL of the necessity of encoding (E) information to register long-term explicit memory (M) and the CCPL of repetition (R) to reinforce retrieval tracks of information. A typical PE in psychology can mix possibilist with CCP laws to mitigate the presumption of impossibility or improbability of a phenomenon. Let us see it formally:

1. $\forall x\ (\neg Ex \rightarrow \neg\Diamond Mx)$ (possibilist law)
2. $\forall x(CPx \wedge \Delta Rx \rightarrow \uparrow Prob(Mx))$. (ceteris paribus law)
3. In S, CPa (the other conditions are supposed constant)
4. In S, Ia $\rightarrow$ Ea (in an S context, I is permissive)
5. In S, Ia (The permissive condition I is satisfied)
6. $\therefore$ In S, Ea (The necessary condition N is satisfied)
7. $\therefore$ In S, $\neg(\neg\Diamond Ma)$ (The memory was not impossible)
8. $\therefore$ In S, $CPa \wedge \Delta Ra \rightarrow \uparrow Prob(Ma)$ (repetition raised memory formation)
9. $\therefore$ In S, $\neg(\neg\Diamond Ma) \wedge (CPa \wedge \Delta Ra \rightarrow \uparrow Prob(Ma))$

We can read the conclusion in plain English as: "In the given context (S), it was not impossible for John to form a new explicit long-term memory (Ma) of the kissing bug, and given that the background conditions (CPa) held and that there was repetition (ΔRa), the probability of forming and retaining that memory increased."

6.4.2.2 Mozart Mastering a Musical Instrument

How can we explain Mozart mastering the clavichord at the age of five? We cannot do it sufficiently and may never be able to. However, we can offer a scientific and reasonable PE.

There are necessary conditions of several causal levels for mastering a musical instrument. At the neurological level, refined motor skills (R) coordinated by the cerebellum are necessary for mastering (M) the keyboard. Mozart had intensive practice (P) since age three, which was permissive for his motor skills. At the informational level, knowing musical theory and techniques (T) is also necessary. His father, Leopold, a renowned music teacher and virtuoso, had ministered to him lessons (L) of music from the early age of three. At the socio-ambiental level, we should recognize that if a child does not have socio-cultural exposure (E) to music until he is five, it is not possible for him to master (M) the clavichord by this age. As we know, Mozart was born in a culturally (C) rich musical environment.

Moreover, if continued and intensive practice (until some threshold) is a permissive condition for developing refined motor skills, it is also true that the more practice above this minimal threshold, the more chances of mastering the instrument. Mozart had practiced far beyond the average and, therefore, had improved his chances of becoming a Master at five. To be clear, the necessary condition of refined motor skills makes it possible, but its permissive condition of intensive practice is also a contributory condition that improves its chances.

From all this, we can say that the fact that Mozart became a master of clavichord at the age of five is explained by being born in a rich musical environment with a dedicated father who, being himself a musical teacher and virtuoso, had taught Mozart since three years old, reinforcing his behavior of daily practice and repetition (sufficient continuous practice—CP) and presenting him the information about musical theory and techniques. This is a good explanation of Mozart, but it still does not explain why Mozart was Mozart. Until now, maybe thousands of children had the same conditions satisfied and did not become anything like Mozart. Still, it explains how some things made him possible.

We can offer the structure of this complex explanation as:

1. $\forall x\,(\neg Rx \rightarrow \neg\Diamond Mx)$
2. $\forall x\,(\neg Tx \rightarrow \neg\Diamond Mx)$
3. $\forall x\,(\neg Ex \rightarrow \neg\Diamond Mx)$
4. $\forall x\,(CP(x) \wedge \Delta Px \rightarrow \uparrow Prob(Mx))$
5. In S, $P \rightarrow R$
6. In S, $L \rightarrow T$
7. In S, $C \rightarrow E$
8. In S, Pm
9. In S, Lm

10. In S, Cm
11. In S, CPm
12. ∴In S, Rm (from 8 and 5)
13. ∴In S, Tm (from 9 and 6)
14. ∴In S, Em (from 10 and 7)
15. ∴ In S, $\neg(\neg\Diamond Mm)$ (from 8, 9 and 10)
16. ∴ In S, $\uparrow Prob(Mm)$ (from 11, 8 and 4)
17. ∴ In S, $\neg(\neg\Diamond Mm) \wedge \uparrow Prob(Mm)$

This formalization combines the necessary, permissive, and contributory conditions that collectively explain how Mozart's mastery of the clavichord at the age of five was possible and made it more probable. Of course, it is not sufficient to explain Mozart.

It is becoming more complicated, as it is psychology. Should I insist on formalizations? I do not and will not. I have always shown them up to this point, accompanied by plain English descriptions of the explanations. The objective was not scholastic but to show that PEs are rigorously logically structured and have clear means of falsification. Moreover, they are essential to demonstrate that explanations can be offered, even in psychology, with the same structure of arguments.

I can complement this description with non-formalized improvements to previous PEs. To the explanation of my insight about an electronic circuit in high school, I can say that in addition to the fact that it was possible because, at the time of the insight, I was in a relaxed and open state of mind, I had satisfied the necessary condition of having a cognitive framework that could integrate disparate pieces of information, and had improved my chances of insight by reflecting more time on the problem than the average time of my colleagues.

About the example of my attainment of a flow state in writing this book, in addition to the fact that I consider this a challenging activity and I am engaged in it, there are other necessary conditions to flow state satisfied, such as being free from distractions and intrinsically motivated to engage in the writing. All the other examples of psychological PEs in this book could also be significantly improved.

What I did here is to show that PEs can be composed, and usually are, and that is why it is not obvious to identify them. We can see here that layers of causality (informational, neurological, behavioral) can be added indefinitely to this kind of explanation, enriching it with more and more necessary and contributory conditions that add, each of them, comprehensibility to the explanandum. This is possible because neither CCPLs nor PLs sufficiently determine the event, which would make any other condition irrelevant to the explanation.

6.4.3 The Limit of Scientific Psychology

Everything we have considered up to this point leads to a foundational conclusion. The realm and limits of scientific psychology are defined by its possibilist and comparative ceteris paribus laws, as well as by PEs (mechanistical included), for the simple reason that only they are legitimate and falsifiable laws in the discipline, and only PEs can be based on legitimate scientific laws and facts.

To illustrate this, let us consider these various sufficient explanations of why Peter runs, trying to find what they have in common. The BDI model could say that Peter runs because he believes that running will improve his health (belief), and he desires to be healthy and fit (desire), which has led to his intention and decision to run. The informational model could say that Peter runs because he has processed the information that regular exercise is beneficial for cardiovascular health, which, through a decision-making process, led him to choose running as a form of exercise. The neurological model could say that Peter runs because of the activation of his motor cortex and the release of neurotransmitters like dopamine and endorphins that create a feeling of pleasure and reward when he engages in physical activity, which led to the strength of synapses that are now prompted activated when receiving specific neuronal signs connecting then to neuronal reactions of the motor cortex that command the cerebellum regulations of the automatic movements of the legs. The evolutionary model could say that Peter runs because, from an evolutionary perspective, physical activity like running was essential for survival, and some environmental signs have prompted his instinctual behavior to maintain physical fitness and readiness. The psychodynamic model could say that Peter runs to deal with unconscious conflicts and anxieties, providing a way to release repressed emotions and offering a sense of control and mastery over his environment. The humanistic model could say that Peter runs because he seeks self-actualization and personal growth, and running allows him to achieve a sense of accomplishment and a meaningful and fulfilling life. The socio-ambiental model could say that Peter runs because his social and environmental context reinforces this behavior, as he lives in a community that values physical fitness.

Did you find out what these explanations have in common? They are not scientific in the strict sense, meaning they do not rely solely on empirical data or corroborated laws or mechanisms. Some of them, as neuroscientific or informational explanations, resemble scientific terminology and are more easily naively accepted as scientific. However, none of them can prove the core of its allegations, because they are explanations of sufficiency. And there are none, not even one sufficiency law or mechanism in psychology, not to mention the unobservable or subjective supposed data in which they are based, even in neurological contexts.

However, at least the models of BDI, informational, neurological and the behavioral components of socio-ambiental explanations can provide possibilist or CCP laws, as well as empirical evidence, to support a possibilist scientific explanation of Peter's running. Nevertheless, every time someone postulates in psychology sufficient causation at any level, he or she makes a reductionist metaphysical option (in the sense that they are setting which is supposedly the fundamental level of psychological causation), at the same time that a scientifically baseless claim.

This would draw a precise frontier between scientific and philosophical psychology. Between science and metaphysics. Nothing against metaphysics, it is just how things are. Science we can test and is falsifiable, metaphysics, until further notice, can only be speculative.

To put it differently. Any claim of sufficient cause in psychology, whether BDI, neuroscientific, informational, evolutionary, humanistic, psychoanalytical or behavioral *must* be metaphysical. Disregarding what any of us consider the fundamental realm of psychological phenomena. But some claims about necessary or contributory conditions to some behavior *may* be scientific.

Illustrating with the case of Julie, who performed a conservation task, the scientific explanation is that having an intact left temporal lobe and having reached a certain stage of cognitive development are necessary conditions for her to be able to understand the request of the task in question, and to solve it. Speculative claims, such as her actions being motivated by a desire to impress her father, or punish her mother, or satisfy her libido, or overcome her basic feeling of inferiority, or fulfil her meaning of life, all fall into the realm of speculative metaphysics.

There is a collateral consequence of this hallmark that is as important as the frontier itself. It is the fact that by identifying the limit of what is possible to know scientifically in psychology with possibilist and CCP laws, scientific psychology becomes totally compatible with an image of a human being that is free, creative and proactive, despite being under the influence of causal pressures and limited by necessary conditions of several orders. At the same time, this hallmark is also compatible with the metaphysical thesis of determinism. The fact that we cannot establish a sufficiency strict law in psychology does not imply that there are no such laws in nature. This limitation could be entirely due to the complexity of the psychological object. In other words, the limit that possibilism can draw sets free psychological science from the metaphysical decision about free will.

6.4.4 A Path toward Unification

This precise frontier can be the new cornerstone to build a new political strategy to overcome the problem of fragmentation of psychology, which is traditionally pointed out as its major obstacle to becoming a mature scientific discipline[15] (Koch, 1959; Staats, 1983; Yanchar & Slife, 1997; Goertzen, 2008; Gaj, 2016).

We should distinguish between pluralism and fragmentation. Pluralism is neither intrinsically bad nor good. It depends on the type we are referring to. I welcome the pluralism of kinds of laws and explanations. Of methods and levels of analysis (such as behavioral experiments, computational models, neuroimaging, field observation). The kind of pluralism that is damaging is ontological relativism, the view that psychological phenomena are entirely perspective-dependent, so that incompatible accounts can all be equally "true." What is problematic is not this methodological pluralism, but the coexistence of incommensurable paradigms that define the study object, epistemic principles, and evidential standards in mutually incompatible ways. A discipline that entertains behaviorism, neuro-reductionism, and social-constructivist narrativism as coequal "theories" is not a coherent discipline; it cannot converge on truth. It has the same event classified as non-existent, neurological, or socially constructed, depending on the paradigm. My position is therefore monistic in its goal, yet pluralistic in its means: multiple methods are indispensable stepping-stones; however, the long-term scientific aspiration remains a unified, coherent, law-guided, and truth-oriented account of psychological phenomena.

Fragmentation is the opposite of unity. A scientific discipline is unified when there is common ground between its practitioners about what existent entities it investigates, which concepts are acceptable to categorize entities and events, which general method should be applied to investigate its object, and which methodologies are better to use in practical situations of research to control and manipulate variables, collect data and test hypotheses. As a result, such a scientific discipline will present a consensus on which practical applications of its knowledge are more effective in achieving a specific result and more unified institutional organizations for its development and dissemination.

From this loose definition of scientific discipline unity, we should expect reasons for its fragmentation. In fact, this is the structure by which I will address the supposed causes of the fragmentation of psychology, which will mirror these components. I view psychology fragmentation less as a political problem than as a derived problem stemming from its ontological, semantic, epistemological, and methodological difficulties.

Royce (1985) argues that all sciences go through four stages of development. First, it begins with pre-scientific philosophical speculations. Second, it starts some empirical exploration. Third, it develops controlled observation and quantification methods. Finally, it achieves formalization and theoretical unification. Considering this theoretical outline, psychology is considered an immature science, as it is, at best, at the third stage, without any formalization or general theory. Mature empirical sciences are physics, chemistry, and biology. So, what can be done to psychology to achieve theoretical unification?

This subitem will not review the extensive discussion on the topic; it is only directed at presenting some of the possibilism's contribution to that. It is divided into three sections that try to synthesize the problem in my terms: the diagnoses of fragmentation (separated between ontological, semantical, epistemic, methodological, and political reasons), the kinds of attitudes and proposals before fragmentation in the discipline, and the path that possibilism can open to help achieve psychology unification.

6.4.4.1 Why Is Psychology so Fragmented?

At the beginning of this chapter, we saw the obstacles that psychology faces in becoming a regular modern science. My basic hypothesis about psychology's fragmentation is that it is primarily derived from the different solutions to or renouncements of these obstacles.

We can clearly identify how the different answers given to these problems give rise to the successive bifurcations that generate the discipline's chaotic picture. Let us observe this summarily with the theoretical framework of the first item of this chapter.

At the ontological level, psychologists are not in consensus about (1) what exists, (2) what we presuppose as existent when we talk about psychological phenomena, and (3) what is the object of psychology. These three problems are the ultimate source of all ulterior fractures in the discipline, beginning with the theoretical. Problem 2 gives rise to the division between those who consider consciousness presupposed by how we describe psychological phenomena and those who consider it an illusion. Problem 3 is the primary source of discipline fragmentation, which can have diverse answers such as behavior, information, consciousness, meaning, essences, cognition, personality, unconsciousness, or neural activity. This diagnosis is widely recognized among psychologists who investigate the foundations of the field, including Staats (1991), Henriques (2011), and Sternberg (2004).

The fragmentation at the ontological level leads to the fragmentation at the epistemological level. Those who choose behavior as the object of psychology consider the old inductive positivist framework enough as an epistemic foundation. Those who think that its object is information processing, cognition, or neuronal activity advocate for a hypothetical-deductive search for knowledge. Those who think essences are the objects, become a phenomenologist, if meaning or unconsciousness, some kind of hermeneutics. Those who opt for an object unsuitable for scientific investigation complain about the cruelty and narrowness of the scientific method and pledge allegiance to supposed alternative methods of the "human sciences" that would be the only ones capable of investigating the true object of psychology. This scenario was classically described in a famous paper by Gregory Kimble (1984).[16]

Of course, in turn, the ontological and epistemological fragmentation leads to semantical fragmentation. Choosing a different object is the primary reason for the semantical chaos in psychology. The second is the diverse conceptions of knowledge that arise from the necessity of investigating different objects. The profusion of theories in this discipline happens not only between versions that adopt different objects and epistemologies but also between those who adopt the same. This incapacity to clearly define the theoretical parameters of this science (Robinson, 1985) produces generalized conceptual confusion (Gaj, 2016) and an "incommensurability" state (Kuhn, 1962), where the translation of one theory into another's terms becomes, if not impossible, very difficult.

Regarding the methodological level, we can only expect that fragmentation at the more fundamental levels leads to a myriad of methodologies that are not recognized across the approaches. Neither experimental psychologists nor psychoanalysts are ready to consider psychoanalytical hermeneutics seriously as a scientific method, nor are psychoanalysts ready to consider experimental series with rats as a psychological approach. Not to mention the measurement artificialism[17] of experimental psychology. Armed with those holy armors, humanists feel justified in vandalizing methodology with several loose and pret-a-portê "methods" for each occasion.

It is only natural that all those fragmentation levels reflect on the discipline's political organization as a human endeavor. However, the political level of psychology adds more layers to this chaos. It is because, usually, institutional psychology rewards the production of novelty, valorizing creativity over paradigmatic construction (Staats, 1983, 1991). This increases the tendency for the proliferation of theories, new terms, and the loose and irresponsible use of crucial concepts. At the same time, psychology has no infrastructure to promote the integration and unification of its knowledge (Staats, 1991; Henriques, 2011; Gaj, 2016). Mature sciences have an entire institutional framework directed to this aim. Another political factor contributing to disunity is the economic necessity of constructing feuds to legitimize each theory's practical applications, primarily psychotherapy. That involves the advertisement activity of denigration of the other approaches. Finally, as psychology uses the same term to describe both the science and the practice of the discipline (unlike biology and medicine, or physics and engineering), the appearance of disunity becomes even more pronounced.

6.4.4.2 Attitudes Facing Fragmentation

Silva (2024) classifies the attitudes facing psychology fragmentation into three great groups: the reductionists, the integrationists, and the skeptics.

The reductionist attitude aims to unify psychology by creating a comprehensive theory, method, or language, or simply by excluding another form of knowledge that is considered inadequate. The most common strategy is to attempt to reduce all other supposed levels of the psychological phenomenon to a fundamental one. Eliminative Materialism, as presented by Paul Churchland (1981), is a typical example of this position. However, a strict reductionist position is not always necessary; the simple attitude of restricting some kind of "knowledge" can be interpreted as a reductionist strategy in the loose sense of the term in this context.

The integrationist seeks the coordination of multiple theories and research results without prior exclusion of any approach. The ultimate goal is to integrate all of them into a comprehensive theoretical framework. Therefore, the focus is on improving communication between traditions through translations, bridging theories, or even by incentivizing the study of psychological phenomena from various perspectives (Sternberg, 2004). Authors such as Sternberg (2004), Staats (1983,1991), and Henriques (2011) are typical representatives of this kind of attitude. However, both Staats and Henriques recognize that there is no possible unification of psychology without a common understanding of the psychological phenomenon, a comprehensive theory, and a shared language, which makes them closer to the "reductionist" attitude that they think.

Finally, the skeptical approach refuses the possibility of any form of unity in psychology, opting to valorize this condition. To them, fragmentation is not a problem but rather a positive sign of disciplinary growth, and we should not pursue unity. Humanists and post-modernists are two typical kinds of psychologists who are used to defend this position. The first kind essentially views meaningful psychology as a human science and believes that modern scientific psychology loses touch with the most critical aspects of psychological life. Moreover, according to authors like Koch (1959), psychology cannot solve the ontological problem; in other words, it cannot avoid different visions about its object or the nature of the psychological phenomenon. Therefore, it is natural for the production of divergent approaches and specializations. When it comes to post-moderns like Gergen (1985), the usual discourse. There is no subject, no object, no truth, no knowledge, no reality, only discourse. Therefore, each identity has its own psychology, and so on.

6.4.4.3 *Possibilism and the Prospect of Unification*

Possibilism is not a theory of unification in psychology; it is an epistemic description of scientific laws and explanations that objectively (but implicitly, until now) exist in science. However, as a missing piece of the scientific psychology puzzle, a very significant one, it can help delineate a better strategy for the unification of the discipline by showing the nature and limits of its scientific knowledge more clearly.

As authors like Staats (1991) and Goertzen (2008) defend, it seems unequivocal that the possibility of unification of psychology depends on the solution or at least the clarification of basic philosophical questions (ontological, semantical, and epistemic), and the most important of them being the nature and methods of the discipline. Although I generally agree with them, I disagree on what is most important. The most important is the most fundamental: ontology. And even though possibilism is not a metaphysical theory, it can help resolve the ontological clash in psychology by addressing the underlying motivation for the dispute at the fundamental level of psychological analysis. It does it because it opens space for scientific psychology to renounce to the pretention of many approaches—such as Behaviorism (Skinner, 1953); Psychoanalysis (Freud, 1940/1949) and neuroscience (Kandel, 1998)—of setting sufficient causes for the psychological phenomena and instead focuses on identifying necessary conditions under which such phenomena become possible.

Most of the fuel for psychological fragmentation is the war over the fundamental level of the mind. Is the psychological phenomenon only cerebral activity that is too complex for our present scientific level and instruments? Or is it the software that, installed in a brain, can perform the functions that we know? Or is there no sense in talking about any of these things, only about manifested behavior? Moreover, what if, as the overwhelming majority of human beings believe, the fundamental level of the mind is consciousness, the one and only thing through which we experience all the other candidates? On giving up sufficient causes, psychology deflates this dispute and sends it to the realm of philosophical beliefs. This enables the establishment of a genuine program of cooperation and translation among all approaches that follow the scientific inquiry method.

How exactly does this magic happen? A materialist neuroscientist, a cognitive psychologist, and a spiritualist humanistic psychologist can easily understand each other, for example, regarding the need for the integrity of the hippocampus to record new memories. The disagreement only surfaces when they should point to the sufficient (determining) cause of recording a specific new memory. Let us go back to John, who remembers a kissing bug in his uncle's garden. No cognitivist nor humanist would be troubled to concede that having a preserved hippocampus was a necessary condition for forming this long-term explicit memory.

In the same way, no neuroscientist nor humanist psychologist would have any problem assuming the necessity of encoding the information about the bug to form it. Finally, neither the cognitivist nor the neuroscientist would disagree that the meaning found in that experience would improve the chances of the boy remembering it in the future. All these allegations could be scientifically tested.

The trouble would only begin if the humanist alleged that meaning alone was sufficient to form the new memory, or if the neuroscientist defends that a hyper-complex, obscure, and unknown chain of chemical reactions imprinted that memory somewhere in the brain that none knows, or will probably ever know.

Nevertheless, if possibilism eventually becomes the hegemonic epistemic description of the kind of knowledge psychology achieves, this second part—the disagreement part, the speculative part of sufficient causes disguised in scientific terminology—would simply disappear from scientific explanations and be treated as it is: philosophical speculation.

That takes us to the field Staats (1991) and Goertzen (2008) believe is the central battlefield of psychology unification: epistemology, the field of the nature of knowledge, and the methods of the discipline. It is exactly here that possibilism sets the basis for the unification of the field of possibilist and comparative ceteris paribus laws, and the methods that test them. Psychology would become the science that reveals the necessary and contributory conditions for human behavior, describes it through probabilistic and causal-computational laws, and tests them through the same scientific method as physics, chemistry, and biology: the experimental method. The same method has yielded the most astonishing results and advances throughout human history.

To implement a successful program of psychology integration, the first step is to renounce any intention to integrate non-scientific approaches into the core discipline. There is no interest in integrating chaotic approaches to scientific psychology; they only work to bring chaos into serious attempts to unify the discipline. We are not interested in integrating all kinds of psychology for the same reason we are not interested in integrating all kinds of physics. Here, my concern is limited to scientific psychology. In this sense, we can say that I am proposing a kind of "epistemical reductionist" strategy to unification.

Setting the scientific method clearly as the method of psychology means that descriptive methods must return to their exploratory condition in scientific endeavors, and that the objective of psychology is to test probabilistic and comparative ceteris paribus laws (or mechanisms) through experiments. This methodologically sets the conditions for unification. All the approaches that can test their hypothesis through experiments are in; all that cannot are out. In this case, "in" means inside scientific psychology, "out" outside scientific psychology, in the field of philosophical psychology.

To limit scientific psychology is not to limit psychology. Psychology, as the rational speculation about reasons, meaning, evolutionary traits, unconscious motives, intentions, and all other dimensions of human existence, will move on as a human science or, as I prefer, as philosophical psychology. Possibilism does not banish these themes, but rather places them in the right context: outside of scientific inquiry. Every time a psychologist in the future attempts to test a feature about them through an experiment, they will do so by translating its terms into those accepted by the core of scientific psychology. So, humanistic, psychoanalytical, and evolutionary speculations will continue as ever, but clearly outside the unified scientific effort. They would be interpretations of the scientific results. However, this unified core of scientific psychology would be humbler, clearly recognizing its limits within necessary and contributory conditions and being open to the complementarity of sufficient speculative conditions about behavior. In this sense, my strategy to unification is also an integrative one.

This limit is set using a plain and straightforward criterion: What uses the scientific method is scientific psychology; what does not use it is not. Therefore, we are referring to neuroscientific, informational, and behavioral levels, as well as cognitive experiments involving constructs that are only employed by BDI explanations. This is very clear but not very simple.

We may now have a clearer understanding of the domain and type of knowledge that psychology can offer. However, we lack (3) a clear domain of its object, (2) the consensual presupposition about which entities we presuppose when we talk about psychology, and about (1) what exists, really. Lacking this consensus, we became lost in a proliferation of terms disputing to represent very close concepts, separated by the previous presuppositions. There is much work to be done in psychology.

The possibilist interpretation of psychology cannot do this work because, as it should be clear now, it is only an epistemological thesis. However, by deflating the dispute about the object of the discipline, it could open space for a compromising solution. Behavior is the consensual source of empirical data in psychology. Neuroscientists compare indirect signs of cerebral behavior (when possible) with manifested behaviors. Cognitivists compare information input with information output through behaviors, inferring the cognitive functions that process it. Behaviorist psychologists relate physical stimuli to manifested behaviors. A consensual object of core scientific psychology would be behavior and cognitive processes, with protests likely coming only from the last remnants of metaphysical behaviorists, those who insist on smuggling unnecessary metaphysical determinist presuppositions into science.

Another compromise solution could solve the second ontological problem. All neuroscientists, cognitive psychologists, and behavior psychologists need to translate their terms and concepts into folk psychology entities. Psychology is applied in practice, and therefore, it does not matter whether a neuroscientist believes consciousness is an illusion or not; what is essential is that regular people believe it is real, and our language assumes beliefs, desires, and intentions as genuine. Therefore, we have here a "bridge over troubled water," a bridge that every approach had to go through to communicate, test, and apply its theories. This bridge, folk psychology terminology, can be used as the point of communication between concepts in an extensive program of unification or translation of terms that would set the conditions for a future general theory of the psyche. The relationship between this conceptual and terminological framework and "human sciences" approaches to psychology would be construed through the same bridge; the exact translation between concepts of, for example, a psychoanalytical theory and scientific psychology would need to point out operational definitions based on neurological, informational, or behavioral terms accepted by the scientific psychology terminology.

Finally, we arrived at the third ontological problem: the question of what truly exists. Does consciousness exist, or is it an illusion produced by cerebral activity? We could also ask if the matter exists, what is "physical," and so on. However, science is not metaphysics. We are pursuing scientific integration here. Therefore, it does not matter the different beliefs about what exists among the various psychological approaches. What matters in this case is only what we presuppose exists when we do scientific psychology. It is a transcendental question in Kantian terms. It has nothing to do with possibilism, but the answer is clear: We presuppose the same things we listed in Chapter 1: (a) there is a real world (realism); (b) we can obtain knowledge about it (epistemological optimism); (c) it submits its functioning and results to the laws of logic; (d) the universe is at least in part regular; (e) the universe is at least in part mathematically describable; and (f) we can represent reality through language (representationalism). That is the compromise solution we must assume in psychology: only what is metaphysically necessary to assume as existent in order to do any science.

Apart from the possibilist epistemic proposal of focusing on necessary conditions, the path to psychology unification outlined here is not detailed and is highly speculative. A detailed and consequent proposal is a work for the future. Maybe, I could dedicate myself to it. However, once again, we can grasp the potential impact of possibilist ideas on real science problems.

6.5 Final Remarks

With this prospect of advancement toward unification, I conclude this comprehensive chapter on the possible impact of possibilism on psychology. I took psychology as the case study for possibilism in science for a couple of reasons. First, because it is the science I know better. Second, because psychology is the core of any possible future modern science of sociology or economics. Whatever solutions we achieve here, we can translate to there.

Chapter 6 comprehensively examines the obstacles that psychology faces in its quest to become a mature modern science. Through a detailed exploration of its ontological, epistemological, methodological, and political obstacles, it is shown how psychology remains a fragmented discipline. This chapter highlights the complexities inherent in psychological phenomena and the diverse foundational problems that contribute to this fragmentation.

I could offer arguments and examples of the impact of possibilism in these problems, as well as offer some proposals for advancements. Equally, I have shown that possibilism can potentially eliminate or dissolve significant remaining obstacles to the constitution of psychology as a modern science.

After this, it was shown what sufficiency deductive-nomological and mechanistic explanations in psychology would look like and why they are currently unavailable. Given psychological multicausality, I made the case that PEs are presently scientific and attainable in psychology. These explanations and laws (along with CCPLs and possibilist mechanistic explanations) may establish the proper boundary between scientific psychology and philosophy, grounded in sufficient predictions and explanations.

The unification of this scientific discipline would take place in the field of possibilist and comparative ceteris paribus laws in the evidence of necessary conditions and causal pressures for certain behaviors; however, never in sufficient conditions for any behavior to occur. It was proposed a realist but non-reductionist epistemological approach for psychology, grounded in modal logic, falsifiability, and experimental method. A non-dogmatic framework that can accommodate pluralism without lapsing into relativism or metaphysical reductionism.

Nevertheless, I could not pass through this chapter without emphasizing that a possibilist framework can dissolve the metaphysical illegitimate exclusion of the possibility of free will from psychology and neuroscience. Possibilism makes free will and scientific explanations compatible without us having to compromise cognitive science with neither free will nor determinism.

On the other hand, a position that insists on maintaining determinist models of explanation of the sufficiency of an explanans to necessitate an explanandum, even despite the evidence of its practical and possibly ontological impossibility, seems, in fact, only to do so to exclude in advance from "scientific psychology" a competing metaphysical belief that is shared by the overwhelming majority of humanity.

Joseph Rychlak (1994, p. 308) believes that if we fail to advance in the work of foundations of psychology, we will see a deepening gap between cognitive sciences and the most basic assumptions of our civilization, which include free will, the importance of intention in action and the relative autonomy of the individual about society. We need to fight the strange and false propaganda that these beliefs are folk psychology not supported by empirical evidence. In fact, all the empirical evidence we have accumulated in 140 years of experimental psychology falsifies the mechanistic and deterministic image of the human being.

With this introduction to metaphysical problems regarding possibilism and determinism, after six chapters of strict epistemological argumentation heavily exemplified, I will indulge myself with a final chapter speculating on metaphysics.

7

Metaphysics and Possibilism

This final chapter addresses metaphysical questions alluded to throughout this book. Its main objective is to evaluate the metaphysical meaning of possibilism in relation to the problem of free will and the laws of nature. This chapter begins by justifying a metaphysical reflection on a scientific subject and substantiating the judgment that compatibilism is false by synthesizing the Consequence and Powers arguments. After that, it analyzes which kinds of worlds could sustain the possibility of free will, investigating the kinds of laws that could govern them, along with the kinds of entities that would be able to be free. Afterward, I will speculate about what would be the case if possibilism were reasserted as a metaphysical thesis, rather than an epistemic one. What if possibilist laws (PLs) are not merely expressions of our limited knowledge but also represent the fundamental nature of the universe? How could such laws emerge from the interactions of innumerable free entities? Finally, we will analyze the metaphysical significance of the fact that there are only two indeterminist types of actual scientific laws in psychology: possibilist and comparative ceteris paribus laws.

7.1 Legitimate Metaphysics of Science

I was hesitant to include this chapter in the book. I feared that I would compromise the fair consideration of an epistemological thesis with the weight of some metaphysical thesis. Metaphysics is where we hide our deepest beliefs, and for their protection, the fiercest prejudice and reactions usually emerge. If someone wants something to be heard, it is wise not to involve metaphysics, the philosophers' religion.

However, I will take my chances. It is important to picture that, depending on the universe's nature, the significance of possibilism could be even greater. Scientific PLs and CCPLs (as expressions of actual laws of nature) do not depend on a possibilist world to exist and function. The world, in fact, can be perfectly deterministic and yet so complex that we can only know the necessary and contributory conditions to a specific effect. But it is not the way around. If the world is not deterministic but possibilist, only PLs can do the job.

After six chapters of firmly grounding arguments on justified and reasonable premises, illustrated by many examples, I indulged myself in this seventh and final chapter with the right to speculate. By "speculation," I mean reasoning that departs from at least one unjustified or stipulated premise to construct a valid argument. It is the "what–if" model of reasoning. It is legitimate metaphysics if we do not present it as something else, such as science.

Nevertheless, overt metaphysical speculation is not the only legitimate form of metaphysics of science. Two other metaphysical tasks are literally fundamental. The first is to do ontology, investigating what entities we presuppose to exist when we (i) use ordinary language to talk about psychological facts and (ii) assume some theory of psychology. As Quine (1948) has taught us, our use of language and the structure of our theories commit us to specific entities, and investigating which they are is paramount to a conscious scientific activity.

The second is to establish the transcendental conditions for the existence of modern science. Collingwood (1940) defines the transcendental strategy as a reverse quest. Once an object or discipline of investigation is defined, we should determine what must exist for the existence and investigation of that object to be possible. This activity is necessary to avoid metaphysical smuggling into science. We should assume as few metaphysical presuppositions as necessary to practice science. The rest, we should investigate scientifically when possible or leave it to the realm of philosophy.

It is important to note the subtlety here. Given a definition, we can deduce what needs to be real for the object of the definition to be real and studied. However, this does not mean that the object of the definition is real or that the study method is possible. It only means that if there is knowledge about this object, these presuppositions must be true.

So, if we define "Modern Science" as "rational investigation with measured and controlled empirical testing of hypotheses about the physical world, to discover its natural laws." The definition neither proves that it is correct from the point of view of the linguistic use of the term (descriptively) nor that it is the best possible (prescriptively). However, if we assume this definition, we can deduce that, for it to be true, we must also assume that, for example, the universe is predictable and mathematically describable.

This chapter does not have the objective of defending a metaphysical thesis. Its purpose in investigating some metaphysical problems is to consider the potential impact that the epistemic thesis of this book (possibilist scientific knowledge) can have. It does not intend to prove that determinism is false or that free will is true, and will not present new arguments for these, even though I consider these questions of fact, not of reason. A question of reason is whether determinism and free will can be true altogether. So, I will revisit this relationship now.

After this, I will speculate about the potential consequences of the hypothesis that laws of nature could be possibilist, not just the scientific laws that result from our epistemic limitations. Here, I will raise more problems than solutions, outlining possible directions for a future investigation of what we could call "metaphysical possibilism."

7.2 Incompatibilism

To begin, however, I will go back a few houses, presenting more robust arguments in favor of incompatibilism. Throughout this book, I presented epistemological arguments for the relevance of possibilism. However, now I want to show that, given some very reasonable and generally believed metaphysical positions, the relevance of possibilism may be even greater than a mere epistemic solution. Let us start by evaluating what possibilism could mean if incompatibilism is true.

7.2.1 Further Clarifications on Free Will

The underlying cultural problem of the conflict between determinism and free will arises primarily because of the metaphysical thesis of incompatibilism: "Determinism and free will cannot be true simultaneously." Of course, the conflict would be solved by refuting this proposition, and indeed, its truth value seems problematic to some who still endorse a compatibilist effort in philosophy.

However, as we did with determinism, the first task of someone who must address the problem of free will is to define it. Not an easy task. Maybe that is why we are going to start working with the definition provided by Fischer (1999), which is that free will requires both (i) genuine alternative possibilities of output and (ii) control of one's own actions. This means that when writing this paragraph, I selected these words from a diverse set of possible options that would make it possible to express this idea effectively, and that I had control over my decisions and actions in doing so.

However, I will defy this definition a little. (i) posits that without genuine alternatives of outputs, there is no free will. However, even when someone has no alternative possibility of output, if they may control their attitude and thoughts about it, intuitively, we know that there is some amount of free will. In the same way, Fischer affirms that it is a necessary condition for free will that someone have (ii) control of their own actions. I must only qualify this condition, restating it as (ii) some control of some of their own actions.

I have no power to fly, and it does not mean that I am not free. Accordingly, if I become paraplegic, I will not cease to have free will because I am not able to control the action of walking. Moreover, we never have complete control over our own bodies, as any soccer player who has kicked the ball out of bounds in a penalty knows well. Our control of anything, even of our own body, is always partial.

Therefore, I will redefine free will as "having (i) genuine alternative possibilities of action and (ii) some control (power) of one's own actions." Is this definition compatible with determinism?[1] Definitely no.

The revised consequence argument (van Inwagen, 1983) and the powers argument (Mumford & Anjum, 2014) are cogent and demonstrate that the possibility of free will (or the exercise of original causal powers) implies a world where not everything is under the rule of necessity. Let us examine these arguments now.

7.2.2 Consequence Argument

Van Inwagen's (1983) consequence argument shows that if determinism is true, our actions are not free because they are the consequences of laws of nature and events in the distant past, over which we have no control. In its new formulation (van Inwagen, 2008), it seems valid and true so far. Disputes about its validity are basically limited to disputes about the validity of the conditional rule (van Inwagen, 2008). To make this argument clear, let us assume it in its intuitive formulation:

(i) Premise 1: Determinism states that everything now and in the future is a consequence of strict laws of nature and past facts.
(ii) Premise 2: No one has control over the laws of nature nor the facts that happened before we were born.
(iii) Premise 3: If no one has control over laws of nature and past facts, and if they determine everything (determinism is true), no one has control over anything.
(iv) Conclusion 1: That includes our actions. If our actions are a consequence of strict laws and past conditions, then no one has control over their own actions.
(v) Conclusion 2: Therefore, if determinism is true, no one has control over their own actions, and therefore, free will does not exist.

Let us recapitulate. The first premise defines "determinism" as the thesis that the state of the world at any given time, combined with the laws of nature, determines the state of the world at all future times. The second premise, self-evident, states that no one has control over events that happened in the distant past or the laws of nature. The third premise, also self-evident, states that if determinism is true, then every action or event is the consequence of past events combined with the laws of nature.

Therefore, the first conclusion is that our current actions are the consequences of these past events and laws. The second conclusion is that if we have no control over the past and no control over the laws of nature, and our actions are the consequences of these factors, then we have no control over our actions.

There is a second conclusion. Since having free will requires having control over our actions, and determinism means we do not have such control, it follows that if determinism is true, we do not have free will. Therefore, a third conclusion derived from the consequence argument, a theoretical one, is that free will and determinism are incompatible: incompatibilism is true.

Let me put some flesh on this skeleton. Suppose you decided to stop reading this annoying book. According to determinism, this "decision" was determined by a long chain of prior events and the laws of nature. If we trace this chain entirely, we will find ourselves in the first seconds of the Big Bang (or even further, depending on what was before the Big Bang). Hence, if determinism is true, your decision to stop reading this "incompatibilism nonsense" was not within your control—it was the inevitable outcome of prior events and natural laws.

In summary, the consequence argument shows that if determinism is true, no one has or has ever had a choice about anything. What we assume to be our free actions are determined by the past and the laws of nature, and as a result, they are not dependent on us. Therefore, they are not truly free. The principle of alternative possibilities states that we are not morally responsible if we cannot do otherwise. Therefore, one of the consequences of the Consequence Argument is that if determinism is true, we are not morally responsible for anything.

7.2.3 Powers Argument

Another strong argument against compatibilism is recent and based on the metaphysics of powers, the "powers argument." For Mumford and Anjum (2014), a power-based solution to laws of nature must be committed to incompatibilism, as the empowerment of agents is incompatible with determinism. It would be presented like this (Mumford & Anjum, 2014, p. 21):

(i) Premise 1: Causal powers are dispositional properties that inherently involve indeterminacy.
(ii) Premise 2: Free will is the exercise of an agent's causal powers that requires genuine alternatives for action.
(iii) Premise 3: Determinism negates indeterminacy by necessitating every action through prior facts and laws of nature.
(iv) Conclusion 1: Therefore, if determinism is true, there are no causal powers.
(v) Conclusion 2: Therefore, if there are no causal powers, free will cannot exist.
(vi) Conclusion 3: Therefore, free will and determinism are incompatible.

Mumford and Anjum (2011) define causal powers as inherent properties of entities that dispose them to behave in a specific way under appropriate conditions. These powers are tendencies rather than strict deterministic causes; they are dispositional. This means that they do not necessitate manifesting in every instance. They depend on the context and conditions for their manifestation. Therefore, defined in such a way, powers need an element of indeterminacy in how and when they manifest.

Regarding free will, to have it, agents must have genuine alternatives for action; therefore, it also implies indeterminacy. In causal powers theory, free will is the exercise of the causal powers of consciousness. However, if determinism were true, every action would be necessitated by prior states and laws of nature, leaving no room for the indeterminacy required by causal powers. Therefore, determinism is incompatible with the existence of free will as it negates indeterminacy. Thus, compatibilism is false.

The best way for a compatibilist to challenge the powers argument is to challenge the premise that free will is the exercise of one's causal powers or to question the definition of causal powers as requiring indeterminacy. If a disposition is defined as a necessary reaction given the presence of some condition and a "causal power" as such a disposition, then determinism would become compatible with causal powers metaphysics. The problem would be, in such a case, how to define free will. Compatibilists would defy this argument, arguing that indeterminacy is not required for free will and that freedom can be understood in terms of acting according to one's reasons and desires. Let us address this problem.

7.2.4 Compatibilism and Free Will

Compatibilism is the belief that free will and determinism are compatible. Its central strategy often involves redefining "free will." This approach is common when attempting to reconcile seemingly incompatible concepts. However, the question arises: does the reshaped concept fulfill the same role as the original? Does it matter?

Philosophers like Dennett (1984) argue that free will should be understood not as the ability to have genuinely chosen otherwise or to generate new chains of causality but as the capacity to act without external constraints. This is the Lockean concept of "negative freedom," which defines freedom as the absence of external constraints affecting one's actions. According to this view, an individual is free if they act and reason according to their desires, which are themselves caused by antecedent conditions and sufficiency laws.

However, this redefinition raises the question: in what sense can this be called "free will"? If an individual's desires are sufficiently caused and beyond their control, they are not truly free. This concept does not align with the traditional understanding of free will, central to our culture, law, and economy, which is rooted in the Judeo-Christian tradition; therefore, it is of limited interest to our discussion. The central debate in psychology and philosophy concerns whether humans can exercise their own causal powers to influence their behaviors despite numerous variables.

Beyond this, Dennett's definition is problematic because it is challenging to conceive of humans acting entirely without external constraints. From genetic inheritance to environmental influences, humans are always subject to various external factors. Therefore, redefining free will in this manner does not address whether individuals have control over their actions within a deterministic framework.

Another famous compatibilist attempt was that of Frankfurt (1971). He distinguishes between two types of desires: first-order, which desire to do or not to do something (drink Cola), and second-order desires, which are desires about desires (the desire for the desire of drinking a juice, instead). A person may have a second-order desire to want to drink healthily and therefore not want to desire to drink Cola. According to Frankfurt, what distinguishes persons from nonpersons (such as animals) is the capacity for reflective self-evaluation, which involves forming second-order volitions. This capacity allows individuals to reflect on their desires and endorse or repudiate them.

Again, for Frankfurt, free will is not about the ability to have chosen otherwise in a metaphysical sense but about having the will (first-order desires) that one wants to have (second-order volitions). An individual has free will if they can align their first-order desires with their second-order volitions (that would reflect their "true self") and act upon them without internal conflict.

This definition of "free will" would clearly be compatible with determinism, as second-order desires would be themselves caused sufficiently by laws plus conditions. But, again, creates another concept that does not matter for our discussion.

Compatibilists argue that these conceptions of free will allow moral responsibility. They claim that actions are morally assessable if they stem from the agent's internal states, such as beliefs and desires of the second order. Despite a deterministic environment, if our actions result from our reasoning, we would be responsible for them. However, if our reasoning is determined, I could not see how it could be the case.

Another common compatibilist argument is that free will involves responsiveness to reasons. An individual possesses free will if they can process reasons and act accordingly, regardless of whether the background conditions are determined. Yet, this does not resolve the issue, if the reasons themselves were necessarily and sufficiently caused.

Furthermore, characteristically, they cannot offer practical lines of action in the face of the political and even legal conflicts that are beginning to be felt between these two fundamental theses of modernity.

In conclusion, the reshaped concept of "free will" from compatibilists does not fulfil the same role as the original. It does not matter for the quest to know if human beings can initiate new chains of causality, create ideas, and behave with some level of independence from physical laws.

7.2.5 Determinism and Explaining Sufficiently

Compatibilism (in the sense of free will that interests us here) is false. Therefore, if we want to preserve the possibility of free will as a hypothesis for scientific knowledge, we must define "science" without assuming determinism as its metaphysics.

However, current laws and mechanism models aim to explain phenomena's behavior sufficiently. Even when they fail (which always happens), they make clear their commitment to a determinist metaphysical creed that attributes their error to our limited knowledge about facts, laws, or mechanisms. That means that if they had all the knowledge, they would have been able to predict phenomena outcomes perfectly. It is the Laplace demon all over again.

I have defended my argument that for the deductive-nomological and the causal-mechanical models of explanation, "a complete scientific explanation of event E must show how E was epistemically necessary." And if it is true, along with the discrete epistemic belief that everything could be scientifically explained, determinism follows: everything is under the rule of necessity. Therefore, there is no free action.

This has been the "mechanism" by which philosophers such as Dennett (1984) or Frankfurt (1971) exclude, by begging the question, the possibility of free will being compatible with the scientific worldview. Assuming determinism, they corrupt the meaning of free will, replacing it with "action without external constraints." This is metaphysical contraband disguised as epistemology or, many times, as empirical results (such as in Libet's experiment, 1983).[2]

Both the "old" anti-metaphysical logical positivist philosophy and the "new" mechanistic philosophy maintain this compromise. It does not matter whether laws or mechanisms govern us; this rule continues to be sufficient for our behavior and everything else in the universe. Again, both "old" and "new" philosophies of science (metaphysics of science) cannot deny their affiliation with a long philosophical tradition that goes back to the Cartesian "machine" metaphor, introduced into biology by La Mettrie in the eighteenth century.

Popper (1979) stated that physical determinism is the greatest nightmare of our time because it posits that the universe is a vast automaton and that we are nothing more than small gears within it, thereby undermining the concepts of creativity and freedom. Moreover, it is surprising that, despite the conversion of quantum physics to indeterminism, it is precisely where determinism has never been fruitful and clashes with our most basic subjective experience—in psychology and neuroscience—that La Mettrie's ancient doctrine of the man-machine has today more fans.

However, as we have seen here, reality appears to be quite different from what determinism claims. First, I have already argued here that determinism is unnecessary for science. Second, considering the results of contemporary physics, determinism is probably false. Moreover, if it is probably false even in physics, what is the reason for imposing it on psychology, where it simply never worked? Where is the number of variables so overwhelming, and there is not even a prediction of a simple behavior?

7.3 Free Will Necessities

Determinism, by definition, excludes the possibility of free will. This poses two questions: What kind of worlds would allow free will? And, slightly different, what kind of worlds would guarantee free will? To answer the first question, we need "only" to examine the options available for the relations between laws of nature and facts. To answer the second question, we should speculate about the fundamental nature of reality's substances. Let us start with the first.

7.3.1 Three Kinds of Indeterministic Worlds

For free will to be possible, we must inhabit an indeterministic world. But what is this, exactly? If determinism is "the thesis that antecedent conditions and the natural laws that govern them sufficiently cause every event" (Castañon, 2021b), the prefix "In" implies only a negation of the proposition. In other words, "indeterminism" would be "the thesis that antecedent conditions and the natural laws that govern them DO NOT sufficiently cause every event."

Determinism and indeterminism are metaphysical positions regarding laws of nature and their relationship with facts. I argue here that there can exist three distinct types of worlds, each characterized by different indeterministic relations, and in two of these, free will remains impossible. Let us examine them, but before this, let us remember briefly Chapter 2 and the properties that a supposed "nondeterministic law" could have.

7.3.1.1 Indeterministic Laws

Merlussi and Cartwright (2018) identified three ways in which a law of nature can be indeterministic. First, a law can be limited in its extent, meaning it does not cover all occurrences within its domain but only applies under specific conditions. Second, a law could be permissive, which means it allows a range of values as valid outcomes. In other words, a permissive law does not predict a single happening but allows some variation in its correct prediction. Third, a law could be unreliable, meaning it does not always work; what it says to happen does not happen all the time.

The most interesting thing to consider about these properties is when we evaluate not a single law but the complete set of laws of nature L. Thinking that way makes it easier to identify which kind of world would allow free will. Merlussi and Cartwright (2018) identified the possibility of L being limited (in extent), permissive, or unreliable as the only ways to an indeterminist universe. To them, these three possibilities are the only possible sources of contingency about what happens in the world. It has come to the point to refute this thesis.

Let us reconsider these characteristics of L. To say that the complete set of laws of nature is limited in extent is to say that there are happenings (like miracles) or whole domains (like human minds) that L does not cover. To say that L is permissive means to say that it has a set of outcomes permitted for each causal event. To say that L is unreliable is to say that the predictions of L can or cannot happen.

Usually, when someone speculates about the possibility of L being limited in extent, they have in mind some specific domain of facts in the universe: the acts of a free agent. In that case, everything in the universe would be ruled by laws except the acts of consciousness. This would open space for free will. It is not an irrational belief, but here, there is a better option available for those who have this kind of concern: a possibilist world. Nevertheless, let us first analyze each option of an indeterminist world.

7.3.1.2 Anarchic Worlds

The first possibility for an indeterministic world can be one with no strict laws or where laws are merely tendencies that fail unpredictably, lacking guaranteed outcomes. Mumford and Anjum (2015) call this "Stochasticism."

They introduce this concept as a fourth option in the debate on free will, alongside determinism, libertarianism, and compatibilism. Stochasticism would be the lack of both determinism and free will, resulting in a world where events occur randomly without any deterministic or agent-based control.

The authors argue that pure contingency (radical indeterminism), understood as the idea that any outcome can follow any event, precludes authorship and control over actions. For instance, if a decision or action results from a random event in the brain, it is difficult to attribute responsibility to the agent for that decision. Similarly, random bodily movements (e.g., spasms) cannot be considered genuine actions of the agent.

I would like, however, to use another name for this kind of possible world. "Stochastic" comes from the Greek word "στόχος" (stokhos), meaning "aim" or "target." Usually, when used in mathematics and statistics, this word refers to random or probabilistic events governed by probability distributions. In a random process, individual events are independent but follow a known probability distribution. Rolling a fair die is an example; each roll is independent of the previous, and the outcome of each one cannot be predicted, though the probabilities of each outcome are known.

However, the kind of world we are discussing here is different than that. Therefore, I will call such a world "anarchic." In an anarchic world, events occur arbitrarily, not by necessity or even by probabilistic chance. In this world, laws are at most "tendencies" without predictable outcomes. We cannot even establish the probability of any event.

Using the properties discussed above, we can say that in an anarchic world, it does not matter if L is limited in extent. If there is such a thing as L in an anarchic world, it can be unlimited and even impermissive, but necessarily it will be unreliable, meaning that, if there are laws, they are arbitrarily and unpredictably applied.

Anarchic worlds are incompatible with free will because if we cannot sufficiently cause anything, including our thoughts, we lack control over our actions and ourselves. This is everything but free will. Exchanging the slavery of necessity for the slavery of arbitrariness does not give us any freedom. If no laws ensure our ability to move our arms with certainty, we lose control over our actions. Without control, we are not free in any sense, since free will requires the ability to act according to one's intentions.

Thus, an anarchic world undermines our capacity to control nature and our actions. According to Mumford and Anjum (2015), true freedom requires a different form of indeterminism that allows for genuine control and reliable causal influence over one's actions. However, there is no dichotomy between determinism and anarchism. There are other possible forms of indeterministic worlds.

7.3.1.3 Probabilistic Worlds

The second of them is a world where fundamental laws are probabilistic. In this kind of possible world, the laws describe outcomes in terms of probabilities rather than certainties, indicating the likelihood of different possible outcomes given some kind of action (or previous fact).

Using the properties discussed above, we can say that in a probabilistic world, L can be unlimited in extent and reliable. However, what matters to characterize it as probabilistic is that L is permissive, allowing a range of outcomes whose probability distribution is determined by its laws.

However, despite being indeterministic, a world entirely governed by probabilistic laws does not seem to free us from slavery. The dictatorship of chance would substitute the dictatorship of necessity or arbitrariness. Our actions would only activate probabilities, and we would lose control of them and their effect on the world anyway. Therefore, there is no free will in a probabilistic world as it is defined here, because actions are not determined by the agent's choices but by probabilistic outcomes. The element of chance replaces strict necessity (determinism) and arbitrariness (anarchism), which does not suffice for genuine free will.

7.3.1.4 Possibilistic Worlds

However, there is a third type of indeterministic possible world, a possibilistic one. In such a world, L is composed of possibilist and impossibilist laws: there are only laws of possibilities and impossibilities. Such a world would allow free will because it restricts substances' possible outcomes or behaviors but does not determine them.

Let us again recur to L properties. In a possibilistic world, L is unlimited in extent, which means that no event or domain in the universe is not subject to laws of possibility and impossibility. L is impermissive because possibilist and impossibilist laws predict events that are strictly impossible to happen given some conditions, and the absence of this event happens all the times the law is instantiated. However, at the same time, it does not determine what happens in the absence of the event. Finally, L is entirely reliable in possibilistic worlds, meaning its laws always happen, with no exceptions (no necessary condition N, no effect E, always). Negating the only options Merlussi and Cartwright (2018) offer for an indeterministic world, yet one that is clearly indeterministic, metaphysical possibilism opens a new path for thinking about free will in a world governed by strict laws.

In this world, the decisions and acts of a human mind do not need to be in a domain outside the reach of the laws. Laws of possibilities and impossibilities can cover them; to say that a human being cannot walk with a severed spine does not say what he will do seated.

Let us compare it with the previous worlds. In an anarchic world, all the outcomes are determined arbitrarily; therefore, if there exists some substance with the potential to be an agent, it would not be able to control its own behavior. In a probabilistic world, no substance would equally control outcomes because chance would determine them all. In a deterministic world, we already know that all outcomes are sufficiently determined by antecedent facts plus laws of nature.

However, in a possibilist world, meaning a world where the actual laws of nature (not only their approximate representations as scientific laws) are only laws of possibility and impossibility, the substances under the rule of these laws could, if they were agents, act freely within the limits of the laws (i.e., the necessary conditions to act).

Note that a possibilist world does not guarantee free will; it only allows it. Granting that some possible world is possibilistic, the existence of free will would depend on the nature of its fundamental components. That is what we will analyze now.

7.3.2 Substances and Free Will

As we can see, we can only speculate about what kind of laws of nature would support free will. We cannot guarantee the existence of free will based on the kinds of laws a possible world has, not even how those laws are produced. Let us examine the first of these two problems, investigating which kind of substance could exercise free will once we have PLs.

7.3.2.1 Cartesian Mechanicism

The most enduring metaphysics of science is mechanicism. We have already seen its brand-new version here. It was originally conceived by Descartes in his "*Principles of Philosophy*" (1644/1984). Mechanicism affirms that the universe functions like a vast machine, with matter composed of extended substances moving according to mechanical principles initiated by a starting energy or force.

Descartes concluded, using the method of imaginary variations for definitions he created, that the essence of all physical substance (matter) is having extension. This means their geometric properties can be expressed in Cartesian coordinates: length, breadth, and depth. According to this view, matter lacks inherent qualities like color, taste, or smell; these are merely perceptions in the mind. Its only objective quality is occupying some space.

Descartes did not say that matter was everything that exists in the universe. There would be another substance, the mind, that had, among other properties, free will. While Descartes maintained a dualistic distinction between the substances of mind and matter, his empiricist followers, such as the early modern philosophers, often focused on the mechanistic aspects of his philosophy, setting aside the dualism of mind and body.

For Descartes, the universe functions like a machine except for the mind. This follows from his definition of matter as a hollow, passive, and inanimate thing that has as its only property occupying a place in space. Every physical phenomenon can be explained through extension and variations in the extensions between extended objects over time: movement. Movement explains the interactions of the universe's parts, much like the workings of a clock. The only thing that could not be explained is the initial energy put into motion in the system, as matter is inert and passive: the "first motion."

This is where God plays His part in Descartes's model. God must impart an initial motion to matter at the creation of the universe and establish the laws governing its transmission. After this act, the universe operates according to these established laws, with the total amount of motion remaining constant (conservation of motion).

Descartes' God plays a necessary but limited role in this universe. God ensures that all physical events can be explained through mechanical causes. This deterministic outlook arises directly from viewing matter as a passive, nonqualitative entity, always subject to external forces.

The God of Descartes is a theological version of Laplace's demon. If laws of nature are God's Will and matter is an inert and passive extension of space, all physical events can be explained solely through mechanical causes: previous states of motion and the laws of nature. We could very well assume the formula *Cartesius minus cogito pares determinismum*.[3]

7.3.2.2 Leibnizian Metaphysics

At the beginning of Modern Science, mechanicism was not the only metaphysics of science available; it was just the winner. Its direct competitor was Leibnizian monadology, which lost the battle for science at that moment in history.

Leibniz was as far as possible from being the "Dr. Pangloss" sarcastically pictured in Voltaire's Fabula. While Voltaire was, indeed, an intellectual without any contribution to science, Leibniz is reputed as one of the greatest scientists who ever lived, the last universal genius, the initiator of the formalization of logic, the developer of the limit, differential and integral calculus in mathematics, the discoverer of the unconsciousness, the absolute threshold and the "just noticeable difference" in psychology, the discoverer of kinetic energy and potential energy in physics, the first theorist of the concept of cell in biology, the creator of the modern calculus machine in engineering and the theorist of the future computer (logical machine) among others achievements.

Of course, I do not cite these realizations to enable some authority fallacy but only to contextualize a man who was everything but naïve scientifically. On the contrary, in my judgment, he was, at that time, the most extraordinary scientific mind of his era, rivalled only by Newton. His metaphysics is not the extravagant whim of an alienated intellectual but a serious, rational attempt to offer a framework of fundamental beliefs about reality compatible with his time's significant volume of scientific discoveries. His ideas are usually relatively unknown in the English-speaking world because he conflicted with Newton about the paternity of differential calculus and his fierce opposition to empiricist philosophy and mechanistic metaphysics.

Leibniz believed that Cartesian physics was a failure because it stems from a completely wrong metaphysics. He argues that extension (*res extensa*) cannot be the essence of matter because it is not sufficient to explain all its properties. Leibniz proves this by showing that extension does not explain inertia, that is, the resistance that a body opposes to movement. As we all know, a force is needed to trigger movement.

If a force is needed to put a body into motion, it is because this body exerts an opposing force on it. Therefore, there is something beyond extension and movement, which is physical, exerting this force. It is the ultimate nature of matter, therefore, a metaphysical principle.

To show that Descartes was wrong, Leibniz proved that what remains constant in mechanical phenomena is not the quantity of movement (mv = mass × velocity), but kinetic energy, which he called "living force" (1998, p. 158). This can be expressed as the kinetic energy formula that we all know today from high school: one-half the product of mass and velocity squared ($\frac{1}{2}mv^2$).

This correction has enormous metaphysical consequences, notably that the actual constitutive elements of reality are active. Only something that actively exerts force can resist force. The matter is something that is active in space, time, and movement and, therefore, must be understood as a "substance," a metaphysical concept that the moderns were willing to get rid of. Leibniz, therefore, concludes that the essence of not only matter but of everything that exists is *activity*. If something is, it has activity of its own.

To name this thing that has proper activity (nowadays, we could say, "has causal powers"), Leibniz reinterpreted the Aristotelian term "entelechy," indicating a substance that has its own determination and inner purpose. Later, his metaphysics evolved to monadology, and the substance of reality became called "monad," a term borrowed from Neoplatonism.

Therefore, to Leibniz, reality consists of monads, which are centers of force and activity. Advancing quantum physics' basic image of fundamental particles in more than 200 years, Leibniz predicted that the indivisible elements of reality would be active, not passive. These centers of force are simple substances called "monads" to indicate their simplicity and unity. So, to him, everything that exists is a simple monad or a set of monads. Understanding the nature of a monad is understanding the nature of reality.

Leibniz's corrections to Cartesian theses about the essence of matter and reality carry other profound implications. They challenge the notion that space (extension) is the nature of bodies, or the *sensorium Dei*, as Newton suggested. Instead, Leibniz posits that space is merely a phenomenon, a mode of appearance of the activities of beings to our consciousness. This mode has an objective foundation in the relations between things. He extends similar conclusions to time. He argues that time is not an existing reality, but a phenomenon with a well-founded basis. Time, he suggests, is a result of the relation of the coexistence of things and the succession of these coexistences. In essence, space and time are not realities in themselves but phenomena that emerge from the existence of other realities: the monads. With these conclusions, he prefigures some of Kant's ideas by a century.

The Leibnizian critique makes the mechanistic view of the world and bodies as "machines" fall apart. The world is not a "great clock" in which God initiated movement. The universe is alive, full of living forces, until the more fundamental of its substances. Everything in the Leibnizian universe has activity of its own.

7.3.2.3 Contemporary Dispositionalism

Contemporary metaphysics has witnessed a rise in dispositionalism over the past few decades. Dispositionalism is, in many ways, a revival of Aristotelian and Leibnizian perspectives on "substance" and "entelechy." It draws on and updates some classical notions from Aristotle and Leibniz, focusing on the intrinsic properties and active principles of entities.

Essentialist dispositionalism opposes mechanicism and its view of natural entities as passive, with external forces driving interactions between them. It resembles Leibnizian metaphysics in picturing the world composed of active fundamental components, filled with intrinsic forces that drive reality. These "dispositional properties," which are intrinsic to entities, ground natural laws rather than letting laws of nature be external to entities. The classic formulation of this idea is provided by Cartwright (1989), who refers to them as "capacities."

These dispositional properties are also known as "powers." They dispose of objects to behave in particular ways under specific conditions. Fundamental entities are not considered merely passive but inherently active in dispositionalism. Their "capacities" or "dispositions" entail specific causal powers, meaning they have the potential to cause particular effects.

Unlike categorical properties, which are static and describe what an object is, dispositional properties describe what an object does or can do. Dispositionalism is "essential" because these properties are essential to the objects that have them. An object cannot lose these properties without ceasing to be what it is.

Mumford and Anjum (2011) defend that there is a "dispositional modality" referring to the inherent tendencies or dispositions of entities to produce certain outcomes under specific conditions. This modality is neither purely necessary nor purely contingent but represents a unique, intermediate state. Dispositional modality cannot be reduced to other forms of modality (like necessity or possibility) without losing its distinctiveness. It is a primitive modality that cannot be fully captured by conditional analyses. To them, they involve possibilities and potentialities. For example, solubility is a property that involves the potential to dissolve in water. They often manifest only in conjunction with other properties. For example, the solubility of a substance manifests only when it interacts with a solvent.

A supposed "dispositional modality" would have to be distinct from necessity and possibility to Mumford and Anjum (2011) because, first, dispositions do not necessitate their manifestations. Even if something is disposed to produce a particular effect, this effect is not guaranteed due to potential interference or prevention by other factors. Second, dispositional modality involves more than mere possibility. While something being disposed towards an outcome means the outcome is possible, not all possibilities are dispositional. Dispositionality implies a certain natural tendency toward an outcome.

I think the present state of dispositionalism is promising, but obscure. There is no clarity about what a disposition is, only about some things that it is not. Well, it is always a start.

However, nothing in the framework of essentialist dispositionalism necessarily entails the existence of free will. For free will to exist, certain entities must be agents endowed with intrinsic causal powers. This implies not merely being governed by automatic "if–then" dispositions, but possessing control over what would consist of this "then." Mumford and Anjum, I think, would likely not oppose this interpretation, given that they do not regard dispositions as metaphysically necessary in all cases.

If a substance always dissolves when faced with a solvent, or if it dissolves in a proportion on some occasions, or if it dissolves arbitrarily sometimes with it and sometimes not, and even if the presence of the solvent allows it to dissolve, in any of these cases, there is no free will. This substance would only have free will facing a solvent if it were able to control and decide whether it would be dissolved or not.

Free will requires control (Mumford & Anjum, 2014) and entities capable of making choices. A relative contingency of its behavior is not sufficient. If a disposition is contingent solely on possibilities or probabilities without control of the "owner" of the disposition, the world would be stochastic in the first sense (possibilities) and probabilistic in the second. The nature of this contingency matters.

On the other hand, if a "causal power" resumes itself in a guaranteed, impermissible, and reliable "if–then" relation, it is nothing more than an instruction in an algorithm, a necessary relation. A universe constituted only by such a "causal power" would be nothing more than a Cartesian mechanical clock.

In short, if we assume incompatibilism and think about free will in a causal dispositionalist framework, in which causes act on us, we will have to have our own original causal powers, unrestricted by any other "if–then" relation that was not possibilist. If (i) causality does not always mean sufficient causality, and if (ii) we have our own original causal powers, and if (iii) we live in an indeterministic world where the laws of nature would act on us only in a restrictive way, and not in a sufficient way, then we have free will.

In this only world, among others where free will exists, the relationships in nature regarding free agents would be captured only by the possibilist and CCPL models of scientific laws and the possibilist explanations (PEs) that I described and evaluated in this book. The hypothetical problem is solved.

Having described how a world with free will should be, I would like to finish my speculations here by reflecting on how it could form laws, how agent substances interact with other substances of nature, and what kind of impact this could have on a metaphysics of psychology and the mind–brain problem.

7.4 Possibilism as Metaphysics

We just saw that, if incompatibilism is true, only metaphysical possibilism allows for free will. It is time to speculate what this kind of world would be in more detail. How could possibilist and impossibilist laws appear and sustain their enforcement? Supposing that all substances in the universe had a level of agency, how could PLs emerge from their interactions? What are these consequences for a philosophy of mind and the mind–brain relation? And for psychology as a science? More than providing answers to those complex questions, the present item aims to initiate an investigation program into metaphysical possibilism.

7.4.1 Raising a Possibilist World

How could PLs be formed or simply exist? Three possibilities come immediately to my mind (none of which is Humean).

7.4.1.1 Three Ways of Raising Possibilist Laws of Nature

The first is through nomic necessity. A PL of nature can be understood as a relation between universals, properties, or attributes. Certain universals necessitate the instantiation of other universals under specific conditions. A PL arises when one universal implies another universal that is a necessary condition for it, which effectively "grounds" it. For example, the universal "multiplication" requires the universal "addition." The universal of "fire" requires the universal of "oxygen." The universal of "orangeness" requires the universal of "redness."

The second possibility is through dispositional essences. Conceived as dispositions, fundamental laws of nature represent the fundamental capacities of substances to behave in certain ways under specific conditions. In this view, they are "if–then" relations, such as "if certain conditions w hold, then x will behave as y." These dispositional properties are grounded in the essential natures of substances, their "causal powers." For PLs, these dispositions must be more specific. They should allow for a range of behaviors of x in the presence of some condition w, without determining an exact outcome. The corresponding formula for this disposition would be "if certain conditions w hold, then x can behave as y." This formulation expresses the potential behaviors that x can exhibit, rather than determining a single outcome.

A third possibility, entirely disconsidered in contemporary philosophy, is based on monadic metaphysics. In this framework, a PL could be viewed as the expression of God's will, regulating the relations between monads. Suppose everything that exists is seen as centers of activity rather than extensions of a mechanistic world transmitting motion. In that case, this free activity must be contained to establish order in the universe. According to this view, natural laws constrain these activities, and they are thus possibilist in nature.

7.4.1.2 Deriving Other Laws from Possibilist Dispositions

We could use causal dispositionalism to posit that the ultimate constituents of reality have causal powers constrained by natural laws. These laws would restrict possible outcomes by preventing certain events from occurring. If a specific condition is not met, a particular result cannot happen.

This possibility was somewhat implied in Mumford and Anjum's (2011) work. While they do not explicitly label laws of nature as laws of restriction, they consider dispositions inherently restrictive. Dispositions inherently involve constraints on entities' activities, given their natural tendencies to interfere. By grounding laws in the dispositional properties of entities, Mumford and Anjum defend that these laws are not external impositions but intrinsic limitations on how entities interact and manifest their powers.

If dispositions were possibilistic, substances would be free within their range of possibilities. Consider extreme examples: the minimal range of choice is between two states. An electron in a magnetic field presents only two states; its spin is either aligned (spin-up) or anti-aligned (spin-down) with the magnetic field. When we measure the spin of an electron, it will collapse into one of the two possible states (spin-up or spin-down) with a certain probability, but we cannot predict which state a specific electron will assume.

The maximum range of possibilities of choice is infinite. Restricting some finite domain of possibilities leaves the quantity of remaining possibilities infinite. This is generally the case with human behavior. For example, a functional spinal cord and motor system allow for walking. However, if Steve can walk, he might also stay still, run, or jump. Suppose he has restricted the domain of choice by a severed spinal cord. In that case, his behavior options remain infinite, including playing the guitar, driving an adapted car, chatting, or watching a movie.

For substances with a wide range of choices, behavior may seem chaotic or free. However, if aggregated in significant quantities, consider what would happen with substances (or fundamental particles) with restricted or even a dual range of allowed behaviors. The bigger the quantity of agglutinated fundamental particles, the more stable they would appear due to their combined behavior. The behavior of a single electron in a magnetic field is unpredictable, but the combined behavior of a billion electrons appears stable, following a normal distribution. This would be a reasonable explanation of the physical stability of macro-objects.

It may be coherent to conceive a possible world in which fundamental particles are agents limited to a few countable behavioral options. When aggregated in large numbers, such particles would exhibit stable dispositional patterns, known as "if–then" rules, that would emerge as fundamental laws of nature. In such a world, where all substances are agents endowed with possibilistic dispositions, the macro-level could display a complex constellation of laws that appear possibilist, probabilistic, or even sufficient, much like the laws we observe in our own world. Perhaps, identical.

This is not an empirical claim, but a conceptual possibility. Suppose that a basic particle can freely choose only between two options. The statistical behavior of a large collection of particles exhibiting such behaviors would approximate stable ratios, such as a 50/50 distribution. That would mimic what we might describe as a probabilistic or even sufficient strict law. The apparent lawfulness at the macro level, then, could emerge from the aggregation of minimal freedoms at the micro level, without contradicting the underlying agentive nature of reality.

7.4.2 Mind Possibilism

What if consciousness is a fundamental force of the universe? We should consider it seriously; after all, this, and everything else we consider, we do with our consciousness. This does not mean another kind of dualism, or the same, but rather the opposite. We could speculate that everything that exists is conscious at some level.

Leibniz was the first to consider this radical idea seriously in Western philosophy with his theory of monads. The nature of the activity that would characterize everything that exists would be psychic activity. According to Leibniz, monads are the basic, indivisible units of reality, each possessing a form of perception. This perspective implies that even the smallest particles possess some degree of consciousness, challenging not only Cartesian dualism but also, more radically, every kind of physicalism.

This does not mean that electrons have a self, memory, and a tortured consciousness of their tragic destiny. Consciousness would come in degrees, as anything else in the universe. The mind state of an elementary particle would be extremely stereotyped, limited to two perceptions that would never constitute a conscious perception (an apperception in Leibniz's terminology).

7.4.2.1 Human Consciousness

The objective of considering this radical alternative is to prepare the reader for a more economical possibility, which is believed by the overwhelming majority of humanity. If not everything in the universe, then materialists may concede that our mind is conscious, and consciousness is not an illusion. It may seem absurd to write such a sentence, but, in fact, it is well known that some philosophers (Dennett, 1991; Churchland, 1981) simply deny the very existence of consciousness (and not only the causal role of consciousness in the universe).

The dismissive treatment by materialists and physicalists of the immaterial hypothesis of the mind does not improve the weaknesses of their own position, which cannot even define "matter" or "physical" (Mazzoni, 2018). This should prompt the reader to engage in critical thinking. Furthermore, the original determinist hope of finding sufficiency strict laws that could generate nomological-deductive explanations in scientific psychology or neuroscience was a gigantic failure. More than one hundred and forty years of experimental psychology and seventy of modern neuroscience have failed to generate even a single strict sufficiency law of behavior.

In the face of this failure, this original materialist impulse was replaced by the rise of causal-mechanical explanations in neuroscience. Giving up on searching for strict laws, recognizing the impossibility of offering future DN explanations, and sufficiently predicting individual behaviors, the option for mechanistic explanations explaining "capacities" left untouched the ideological core of conventional science in psychology and neuroscience: physicalism and determinism.

I have offered strong arguments to support the idea that scientific psychology can only offer us possibilist and CCP laws. More than that, I have demonstrated that these laws are indeterministic. We should consider, however, that they are not due to our epistemic limitations. After all those years of "promissory materialism" (Popper, 1977, p. 96), we have the right to consider that only such laws of nature govern psychological phenomena. And we have the right to consider it based on empirical evidence. Moreover, we do not have any scientific, epistemic, or empirical motive to think otherwise.

To everyone who has some commitment to empirical evidence, the failure in offering any sufficient law or detailed mechanism able to predict or even explain sufficiently an individual's behavior should account for the free will hypothesis. Although under-determination alone does not prove it, persistent predictive failure raises its prior probability.

7.4.2.2 The Correlation between Nature and Civil Laws

It is curious to notice that the real laws of psychology are correlated with another set of laws designed to exert some control over agents: civil laws. Dealing with (supposedly) free agents, civil laws do not dictate how we should behave today, tomorrow, or in our lives; they dictate how we should not behave. It is a good way to illustrate, analogically, the PLs in nature.

Civil laws usually do not have the imperative form "if x takes place, then s should behave in a way y," but rather constraining "if x takes place, then s cannot behave in a way y, under penalty w of law." That is, if, as defended by causal dispositionalism, the ultimate constituents of reality have causal powers (or, as Leibniz wanted, are foci of activity in the universe), the nature of the laws governing the universe should be constrictive (forces constraining or blocking the result of the activity of a basic constituent) exactly as civil laws. Simply because, dealing with free agents, there is nothing else to do.

The difference from the analogy is that in the case of PLs, the constriction is total: in the absence of the specific conditions, a phenomenon is impossible. In civil law, they are not impossible; they provoke penalties that have the effect of causal pressures to diminish the incidence of undesirable behavior. For example, a citizen cannot kill if it is not for the sake of their own life or the lives of those under their responsibility. To protect his own life, however, it is allowed. But if he wants, he can become a serial killer. The prohibition only acts to exert causal pressure on his behavior. Against the PLs of nature, his will cannot do anything.

This kind of law would be, in this case, the only kind of strict law available in the universe. Despite the subjective evaluation of the plausibility of this metaphysical hypothesis, it is a possible one; therefore, it is just one more reason to have within our models of laws at least one that can express this kind of relation.

7.4.2.3 Mind–Brain Relation

There are two primary hypotheses regarding a mind independent of the brain: Cartesian substance dualism and monistic panpsychism. In substance dualism, the mind is a distinct substance from matter, challenging the principle of causal closure of nature. In panpsychism, the mind is a fundamental constituent of nature, integrated into its causal relations.

Both hypotheses must be excluded from a Laplacian world because if nature is causally closed and every event is necessary, an immaterial substance would be unable to interact with the brain causally. A basic constituent of nature would have its behavior entirely determined, like any other.

A critical distinction often overlooked by materialist monists is between substance dualism and property dualism. Some may believe in only one substance (matter or consciousness) but acknowledge that physical properties cannot fully explain mental properties. Moreover, physical and cerebral terminology cannot even describe mental properties and experiences.

Therefore, a materialist could sustain that there is only one substance, matter, but also that this substance has two properties: physical and mental. Generally, a property dualist defends that mental properties like consciousness or qualia "emerge" from the physical activity of the brain but are not reducible to physical properties. In other words, to him, mental states depend on physical states but are not reducible to them.

Similarly, a panpsychist is a monist who believes that consciousness is the only substance in the universe. However, he obviously recognizes that matter and consciousness are phenomenally different. First-person experiences of consciousness are ineffable mental experiences, while third-person perceptions of other consciousness involve sensory alterations of our mind. That would be "matter": the experience of the causal pressure from another consciousness.

Assuming a kind of property dualism, at least linguistic, is unavoidable in psychology. This is because we can even have conscious experiences of our brain's physical behavior through a PET scan screen, or vice versa. After all, our feeling of fear is phenomenally quite different from the electrical impulses in our limbic system.

Going further, what does the truth of the hypothesis that actual laws of psychology are possibilist or CCP may suggest? Most of our psychological experiences are not limitations, but rather causal pressures that influence the way we think or behave. So, I propose the thesis that the brain acts as a "focalizer" of the conscious and unconscious mind, not a "filter" nor a "producer." It is predominantly accomplished through attentive processes, notably through the activation of the prefrontal cortex. However, various other sensorial processes are constantly focusing the powers of our mind, including the visual system, the auditory, tactile, olfactive, thermosensory, proprioceptive, gustatory, vestibular, and other systems, as well as all the pre-activated memories and the unconscious monitoring of our immediate environment. Running altogether, these processes consume mental capacity and probably make it very unlikely for extraordinary conscious experiences to occur.

This hypothesis is currently completely speculative. Future research will investigate it in detail and compare it with contemporary neuroscience findings. But this is entirely another story.

7.5 Final Remarks

This final chapter has analyzed metaphysical implications of possibilism for the problem of free will and the nature of scientific laws, showing that, depending on what the universe really is, its significance could be greater.

It shows how compatibilism "explains away" free will and why, considering free will the exercise of causal powers within genuine alternatives, it is incompatible with determinism. It was done by synthesizing the Consequence and Powers arguments.

From there, we went out on a speculation about what the kinds of laws and entities in the universe should be if free will is true. The analysis revealed that anarchic and probabilistic worlds are also incompatible with the exercise of free will. Anarchic and probabilistic worlds undermine control and authorship by introducing aleatoriness and randomness.

Instead, I concluded that a possibilist world, where laws are characterized by possibilities and impossibilities rather than deterministic sufficiency, would allow free will. We also saw that to causal dispositionalism entities possess intrinsic causal powers that allow for genuine alternatives without necessitating specific outcomes. This middle ground between necessity and pure contingency provides a coherent basis for free will, aligning with the principles of what I called "metaphysical possibilism."

We considered the broader implications of these ideas for understanding consciousness and the relationship between the mind and the brain. By considering that consciousness may be a fundamental force of the universe, we saw that, even so, PLs would still be able to govern psychological phenomena and the interactions of free agents.

Summarizing, this chapter has argued that for free will to exist, the universe must be governed by PLs that allow for genuine control and alternative possibilities. While these mere exploratory speculations raise many questions, they also set the ground for future investigation into the nature of free will, consciousness, and the fundamental structure of reality.

However, possibilism, even in its metaphysical form, does not imply that consciousness is independent of the brain with causal powers, but it is compatible with this idea and with current knowledge in psychology and neuroscience. In being so, it opens space for the end of modernity's central cultural fissure between the concept of free will and the scientific worldview.

Moreover, when conceived as a simple epistemic thesis, possibilism makes it unnecessary to decide, in the name of the nature of science, for any position regarding the mind–body relationship. This relation can be dependent or independent and begins to be treated as it always should have been: as an empirical question rather than a matter of principle. The good news is that through metaphysical possibilism, we can conceive what an orderly, law-governed, scientifically knowable, and indeterministic universe would look like.

Conclusion

This book presented a systematic framework for possibilist knowledge. It shows that modern science traditionally focuses on knowledge of what is supposedly either physically necessary or probable, often neglecting the realm of what is possible or impossible. Here, it rigorously details how possibilist knowledge can be integrated into scientific inquiry, showing that it can be systematic, formalized, empirical, objective, and falsifiable.

It introduced possibilist laws (PLs), which state necessary conditions for the possibility of events, and impossibilist laws, which state absolute impossibilities in some domain. Clarifying decades of debates over "how-possibly explanations," it formalized possibilist explanation (PE) into a deductive-nomological model (DNM) form, reviving DN logic in a necessary-condition variant.

PEs elucidate how apparently impossible or improbable phenomena could occur under specific conditions. It presented the concept of permissive conditions to solve logical problems and reveal the true structure of an explanation of a possibility. Similarly, to clarify the phenomenon of negative causality in mechanistic explanations, the concept of "preventive condition" was introduced. These tools expand the scope of scientific inquiry, offering a deeper understanding of complex phenomena.

Despite the metaphysical position one might defend about causality, laws of nature, strict laws, or free will, the fact is that we will not always (if ever) be able to make sufficient or probabilistic predictions in the human sciences. Scientific knowledge is not incomplete just because of our methods and empirical evidence, but because of the type of prediction we can make of a phenomenon. Sometimes, we can predict its occurrence. Other times, we can only predict the likelihood of its occurrence. Moreover, even if that is unavailable, we can sometimes predict its nonoccurrence. This is something useful. Furthermore, this is sometimes the only scientific knowledge we can achieve about something, certainly due to the extreme complexity of social and human phenomena and possibly due to the nature of reality itself.

Unlike deterministic or probabilistic knowledge, possibilist knowledge focuses on the potentiality and constraints of events, making it particularly relevant in complex and multicausal domains. This book demonstrated the ubiquity and empirical relevance of possibilist laws and explanations through diverse examples from physics, biology, history, geography, sociology, political science, economics, psychology, and neuroscience.

Possibilism is not only an abstract thesis in a philosophical fight. PLs can be used to make valuable predictions, which could help promote wellbeing and prevent illnesses and harmful situations. We cannot produce a phenomenon using possibilist knowledge. However, we can produce or prevent the (necessary) conditions for its possibility. We still do not know what causes a person to get out of depression. However, if we know a necessary condition not to become depressed, such as a certain level of some neurotransmitter, we would be able to produce a (necessary) condition without which a person cannot get out of depression.

On the other hand, we still do not know the sufficient cause of a disease like Herpes Zoster. However, as we know, it is due to the reactivation of Varicella-Zoster within a human organism (i.e., having Varicella-Zoster within the body is a necessary condition for this disease), we can define the permissive conditions that promote this reactivation. (in an S context). It is by avoiding these permissive conditions that we can block the disease. Even better, by the right means, in principle, we can completely eradicate this disease from the Earth by eradicating this necessary condition, the virus (in fact, this is precisely what we did with polio types 2 and 3).

Psychology served as a case study to illustrate the profound implications of possibilism. By addressing ontological, epistemological, methodological, and political obstacles to its constitution as a mature modern science, we demonstrated how PEs and laws can help mitigate old problems of fundamentals and offer a new path to unify the discipline and advance its scientific rigor.

Risking raising arms against possibilism, I present some speculation about its metaphysical implications, particularly concerning free will and laws of nature. I argued that PLs provide a viable framework for free will, offering a middle ground between necessity and pure contingency. This approach supports genuine control and alternative possibilities, compatibilizing causation and agency.

These consequences show how possibilism can also address a central cultural conflict: the perceived incompatibility between determinism, free will, and the foundations of modern science.

The spirit of Enlightenment modernity is fundamentally contradictory. Our entire political and social system, most economic theories, the world's largest religions, and even our most ordinary social relations assume that human beings are endowed with some freedom of choice in the face of the biological, social, and psychological conditions to which they are subject.

At the same time, the influence of modern science is immense in contemporary society, guiding everything from public policy decisions to individual health treatment choices. And it does so while promoting a deterministic worldview, where all aspects of nature are governed by causal relationships that are sufficient to explain the occurrence of any phenomenon.

Nevertheless, this clash between destiny and liberty is not a product of the Enlightenment alone. It predates even Greek Stoicism and Platonism, originating from diverse religions and beliefs. Even contradictory religions have attempted to reconcile these two strong and incompatible intuitions: that everything is subject to forces beyond our control, and that we have agency over our own behavior.

This conflict is at its apex today. In the past twelve months (writing in July 2025), two books titled "*The Science of Free Will*" have been released. The first, by psychologist Roy Baumeister (2024), presents free will as a product of brain evolution and denies any utility of determinism for psychological science. The second, by physicist Samir Varma (2025), seeks to reconcile Laplacean determinism and free will through the absurd definition of free will as "the absence of being able to predict what will occur" (p. 31); he calls this "free will in practice." For him, if we can never predict how a human being will behave, that is equivalent to having free will. This underscores how critical the issue has become for contemporary society.

In *Possibilist Knowledge*, a coherent framework is presented to reconcile the idea of a natural world governed by rules with the concept of a social world where subjects are agents endowed with some degree of freedom. And it does it without any compatibilist claim or giving up the traditional conception of free will. Few things matter more than this.

The belief in free will enables a person to be held responsible for their actions, whether good or bad. We are not discussing here the mere concept of freedom as the absence of external coercion, but rather the concept of the individual as the initiator of new chains of causality. Without free will, the concept of "responsibility" has no meaning, and someone's current actions become attributed to the general configuration of physical, chemical, biological, social, or even (deterministic) psychological events. Both our legal and moral judgments cease to make any sense.

We can already see some consequences today of the erosion of the concept of "responsibility." On March 11, 2012, US Army Sergeant Robert Bales left his base in Kandahar, Afghanistan, and slaughtered sixteen Afghan civilians, including women and children (The World, 2016, July 30). In his martial court, his defense team argued that neurological factors determined his mental state and behavior at the time of the killings. His defense alleged that his actions were caused by severe post-traumatic stress disorder, as well as by the consequences of a traumatic brain injury that affected his impulse control and decision-making capabilities. Moreover, the defense alleged that Bales had been drinking alcohol and using steroids on the night of the massacre. All considered, these neurological conditions determined his behavior.

This line of defense was not wholly successful but spared Bales from the death penalty. However, strictly considered, these arguments were perfectly in line with the human image projected by contemporary neuroscience or even compatibilism, since, if they were right, he would not have been able to behave differently. How long until this kind of argument is utterly successful in court?

The dominant strain of contemporary neuroscience supports and disseminates the conception of the human being as an automaton. It is an inevitable consequence of its deterministic view of nature. As said once Jerome Bruner (1990, p. 32), the problem with this kind of "psychology of nothing but" (nothing but reflexes, or associations, or electrical impulses) is not that psychologists believe in it, but that one day everyone, from the postman to the prime minister, will come to believe in it. I believe that if this day were to come, our political, legal, and economic systems would collapse. Is there another metaphysics of science available to neuroscience? Time to look for it.

By refuting the necessity of determinism for scientific inquiry and presenting PLs as a robust alternative, I proposed an epistemic solution that preserves the integrity of the modern world's scientific and moral frameworks. This reconciliation allows us to maintain our belief in individual autonomy and responsibility while embracing the rigor and advancements generated by modern science.

The conclusions drawn in this book open new avenues for research in all fields of science and philosophical exploration in metaphysics and philosophy of mind. Possibilist knowledge does not compete with deterministic or probabilistic principles; it complements them where sufficiency or statistical regularity is not available. It offers a coherent middle ground in the long-standing debate between fate and freedom. I hope this book helps us to develop a more comprehensive view of knowledge, one that encompasses more of the spectrum of possibilities in the universe, providing tools to explore the neglected but infinitely fertile domain of what is merely possible.

Annex

Why Is the Formalization of Possibilist Explanation Here Different from That of Erkenntnis?

In Castañon (2021a) I introduced possibilist explanation (PE) with the scheme below:

Let a be the concrete action/event that occurred in the context S.

1) $\forall x\ (Nx \leftrightarrow \Diamond\ Ex)$	(possibilist law, biconditional)
2) In S, $P(a) \rightarrow N(a)$	(P is a permissive condition in a context S)
3) In S, $P(a)$	(P obtains)
4) $\therefore N(a)$	(N is a necessary condition)
5) $\therefore \Diamond\ Ex$	

In this version, I represented the possibilist law (PL) as a biconditional. The objective was to deduce the possibility of event E at the end. The paper explained, however, that we should read the biconditional as a conjunction and that it is the second part of the conjunction ($\forall x\ ((Nx \rightarrow \Diamond Ex) \wedge (\neg Nx \rightarrow \neg \Diamond Ex))$) that makes the PL falsifiable. If E occurs while N is absent, the law is refuted.

In fact, what the PL predicted that could be tested was not the possibility of something but the impossibility of the occurrence of something without its necessary condition. The structure of a PL prediction was:

1) $\forall x\ (Nx \leftrightarrow \Diamond\ Ex)$ (possibilist law)
2) $\neg Nx$ (there is no necessary condition satisfied)
3) $\therefore \neg\Diamond Ex$ (there is no possibility of the effect)

I changed this formalization by a simple material implication:

$$\forall x\ (\neg Nx \rightarrow \neg\Diamond Ex)$$

The change has two motives. First, parsimony. The shorter conditional serves in both prediction and explanation. Second, causal direction. The biconditional erases temporal order, while the conditional keeps the asymmetry: satisfaction of the necessary condition is prior (in time or ontology) to the possibility of E.

This changed what can be deduced in a PE. Instead of the epistemic possibility of E, what can be directly deduced now is the negation of the negation of the possibility.

(i) $\forall x\ (\neg Nx \rightarrow \neg\Diamond Ex)$ (possibilist law)
(ii) In S, $P(a) \rightarrow N(a)$ (in S context, P is permissive)

(iii) In S, P(a) (The permissive condition P is satisfied)
(iv) ∴ In S, N(a) (The necessary condition N is satisfied)
(v) ∴ In S, $\neg(\neg\Diamond E(a))$ (The event was not impossible)

Step (v) yields the removal of impossibility; turning $\neg\neg\Diamond E$ into the positive claim $\Diamond E$ would require an additional premise (e.g., "there is no defeater for E's possibility"). It highlights a key aspect of PEs: they are not about proving absolute (physical) possibility but rather about establishing the plausibility of an event given the available knowledge. If we don't have information indicating that any necessary condition is unmet, we cannot rule out the possibility of the event.

Notes

1 Possibilism

1 The following categories of possibility are set aside for the reasons given below:

1. Analytic possibility. Because there is no agreed catalogue of analytic truths (Quine, 1951) and many identities once thought analytic (e.g., water is H_2O) turned out to be necessary *a posteriori* discoveries (Kripke, 1980), most contemporary modal theorists dispense this category. Hale (2013) notes that once genuine meaning-truths are treated as logical axioms, "analytic possibility" collapses into ordinary logical possibility, rendering a separate category superfluous.
2. Deontic modality. Deontic claims concern what is obligatory, permissible, or forbidden. Because they address normative rather than natural constraints, they fall outside the scope of the present investigation.

2 There is a debate about whether what is physically necessary could have been another way. Many philosophers believe that the laws of nature could have been different; we name them "contingentists." If it is not possible, as think the "necessitarians," metaphysical and physical necessity and possibility would have the same extension. Nevertheless, as we can imagine other possible laws of nature, we have two different concepts. They will be used depending on one's theory of laws. If one holds that the same substances could exist under different contingent laws, a gap remains. As I adopt a dispositional-essentialist view in which laws are grounded in the essential powers of things, I believe that once those essences are fixed, the associated laws are fixed as well. Hence metaphysical possibility coincides with physical possibility for my purposes, and only the latter category will be employed throughout this book.

3 See Section 5.2.1 for a full treatment of Ohm's law as a paradigmatic possibilist case.

4 The single-observation falsifiability test is detailed and justified in Section 2.5.3.

5 An example of the distinction between DNEs and HPEs:

DNE: "Why do giraffes necessarily have long necks?"

"Because natural selection favored individuals with longer necks due to competition for high leaves, as per Darwinian theory."

HPE: "How could giraffes possibly have evolved long necks?"

DNE: "It could be like a gradual process where slight neck extensions over generations allowed better access to food, leading to reproductive advantages."

The DNE provides a necessary explanation (law-based and deterministic), whereas the HPE, as used in Biology, only offers a plausible scenario. It is a possible explanation for something.

That is not the kind of HPE defended in this book, nor was it Dray's. A possibilist explanation is not merely a putative explanation. It is *an actual explanation of an actual possibility*.

6 To disjunctively define a necessary condition according to Reiner (1993) is to list every alternative circumstance that, individually, would satisfy that condition. In effect, one

supplies an exhaustive disjunction—$A_1 \lor A_2 \lor \ldots \lor A_n$—each Ai being sufficient on its own to meet the requirement.

2 Laws and Possibilism

1 The gravitational constant is: G ≈ 6.67430 × 10^−11 m^3 kg^−1 s^−2, where ^ means that the next number is a potency of the former, m is meter (cubic meters with the potency), and s is second (inverse square with −2 as potency).
2 Again, where F is the gravitational force between two objects, m_1, and m_2 are their masses, r is the distance between their centers, and G is the gravitational constant.
3 iff = if and only if.
4 In relevance logic, the formula A →r B should be read and understood as "A relevantly implies B" or"A implies B with relevance"The subscript "r" distinguishes it from material implication in classical logic.
5 At relativistic or quantum scales Newton's formula is replaced, but within its domain it serves as a paradigm of an SL.
6 Maybe it would be good to offer the reader a little glossary of the specialized symbols used in the modal/probability discussion:

□A	used to express epistemic necessity
◊A	epistemic possibility
¬◊E	epistemic impossibility (according to our knowledge, it is not possible that E)
□¬E	another form to express epistemic impossibility
P(E)	probability of E, classical probability function
P(E/C)	probability of E given C
C(x)	C-predicate of x
N(F, G)	"F necessitates G." N-relation F to G; Armstrong-style necessitation between universals

7 Which, in fact, seems to have happened at a residual level, as we can read in Kornack and Rakic (2001).
8 "Permissive condition" is a new concept that will be introduced in the fourth chapter. It is a condition that sufficiently causes a necessary condition for some phenomenon.

3 Explanations and Possibilism

1 This section builds upon my earlier formulation in Castañon (2021).
2 I will not include this presupposition in the formalization because that would conflate two different kinds of possibility: epistemic possibility and doxastic possibility. That would only add one more level of complexity to our representation, potentially leading to the reader's confusion between these two distinct modalities. Therefore, I will keep it simple, relying on the reader's ability to understand that a scientific explanation always refers to what we know, addressing an incongruence between two or more of our beliefs.

3 This formalization is an improvement to that presented at Erkenntnis (Castañon, 2021). An explanation for the necessity of this improvement is presented in Annex 1 of this book.

4 Mechanisms and Possibilism

1 *The Dappled World* is a classic of this approach.
2 A part or component to whose function a given mechanism has no other way to perform.
3 In a stable causal chain A→B→C→D, a causal link is any direct connection between nodes (A→B, B→C, or C→D).
4 Causal pressure refers to the dispositional tendency or force exerted by a causal factor (or power) toward producing or preventing an effect, without guaranteeing it. It operates in degrees, can interact with opposing pressures, and contributes to the outcome depending on the balance or conflict among multiple such pressures or vectors.
5 In other words, valacyclovir is only a contributory (the contribution for the outcome of a contributory condition can be positive or negative) condition because it exerts pressure toward diminishing the virus, but it cannot render the virus's existence impossible by eliminating a necessary condition for it (i.e., it is not preventive).

5 Sciences and Possibilism

1 An orthogonal pure state is an entirely pure way of describing a quantum system. Two states are orthogonal if they are mutually exclusive. That means they are like two directions at 90°, completely separate from each other. Two pure states A and B are orthogonal if their product is zero, that is, if a system in state A has zero probability of being found in state B, and vice versa.

6 Psychology and Possibilism

1 For more detailed considerations of most of these obstacles, see Castañon (2009).
2 The null hypothesis (often written H_0) is a default claim that "nothing systematic is going on. Usually it is formulated as "There is no difference between the variable values of Group A and Group B." In statistics, you collect data and calculate a p-value to see how surprising that data would be if H_0 were true. A small p-value (typically < 0.05) leads you to "reject" H_0. What Meehl points out is that with large enough samples, even tiny and uninteresting effects will make we reject H_0. So "$p < 0.05$" does not tell us that a law-like regularity exists. Rejecting a null like "$m_1 = m_2$" usually just means some difference exists, not that our theory's specific prediction is right. A good scientific law should face the chance of being clearly wrong (Popper). Because the null is almost always false in psychology (human behavior is rarely identical across conditions), rejecting H_0 is too easy; it doesn't put the theory in real danger.

3 The cited essay from 2000 is reprinted in Cummins (2010), *The World in the Head* (pp. 282–310). Oxford University Press.
4 Note that, as Fodor has argued, this is a problem only to psychophysical laws (bridge-laws), not pure psychological laws.
5 The degree to which the conditions and results of psychological experiments reflect real-life situations.
6 The entire formula is $\forall x\ ((H(x) \wedge (E(x) \wedge \neg I(x))) \rightarrow L(x))$.
7 It is important to declare that this discussion is illustrative and hypothetical. Nowadays, there are reports of cases (Zotey et al., 2023) of severe hippocampal damage, where some degree of functional recovery has been observed due to the brain's plasticity. However, the extent of recovery can vary greatly depending on the individual and the specific nature of the damage.
8 Kaspar Hauser was a man discovered in Nuremberg in 1828 with minimal speech abilities. Supposedly, he had lived his entire life in isolation in a basement with chickens. His late acquisition of language became a famous case study in linguistics and psychology, highlighting the critical periods for language development. However, many aspects of his story (including his alleged noble lineage) have been discredited by historians and researchers. Still, it remains an illustrative tale.
9 The perforant pathway (via perforans) is a critical neural tract in the brain's hippocampal formation.
10 A good survey of this new approach could be found in R. Gifford (2014). *Environmental Psychology: Principles and Practice* (5th ed.) (Colville, WA: Optimal Books).
11 To Jerry Fodor, "intentional laws" are generalizations that involve propositional attitudes such as beliefs, desires, and intentions. They are typically formulated in folk-psychological terms. These laws express patterns like: "If a person desires X and believes that action A will lead to X, then they will do A." To him, intentional laws describe the regularities of mental states and behaviors but are not strict laws, rather, they are hedged by ceteris paribus clauses ("everything else being equal").
12 Classic of neuroscience are the pioneering studies of Hubel and Wiesel (1962), which started from single-cells recordings in cats, leading to the discovery of simple and complex cells in V1 that respond to specific orientations and movements of light bars.
13 The examples used in this paragraph are adapted from Castañon (2021b).
14 An explanation based on the will of the agent.
15 In 2004, the American Psychology Association released *Unity in Psychology: Possibility or Pipedream?*, which is a hallmark of this discussion organized by Robert Sternberg with the contributions of some of the most relevant theoretical psychologists at that date, such as Gardner, Rychlak, Kimble, Staats, and Koch.
16 G. A. Kimble (1984). Psychology's two cultures. *American Psychologist*, 39(8), 833–9.
17 The tendency to create highly controlled but unnatural measurement contexts that distort or oversimplify psychological phenomena in the name of precision.

7 Metaphysics and Possibilism

1 The thesis that antecedent conditions and the natural laws that govern them sufficiently cause every event.

2 The famous Benjamin Libet's experiments on the readiness potential, published in 1983, are a classic example of how the metaphysical belief in determinism is smuggled into science. In this study, Libet measured electrical activity in the brain (readiness potential) that precedes voluntary motor action (like flexing a wrist). Participants were asked to note the exact moment they consciously decided to move using a clock. Libet found that the brain activity predicting the movement occurred several hundred milliseconds before the participant was aware of their decision to move. From this, several determinists concluded that unconscious brain processes initiate actions before the conscious intention, which was sufficient to spread the gospel: "There is no free will." This conclusion, based solely on this experiment, is patently absurd. Asking participants to report the moment of decision based on a clock is subjective and imprecise and cannot reliably track conscious intentions. When it comes to being a milliseconds issue, it becomes a bad joke. Moreover, Libet's task (wiggling a finger or wrist) was too meaningless and simple to generalize to morally or personally significant decisions. The attributed significance of the readiness potential is even more problematic; in the real world, it does not seem to reflect a decision at all but rather fluctuations in neural activity that become relevant when the subject is inclined to act. All considered, we can see that there is little science and too many personal beliefs involved in these deterministic conclusions.

3 Descartes without cogito equals determinism.

References

Alberts, B., Johnson, A., Lewis, J., Raff, M., Roberts, K., & Walter, P. (2014). *Molecular Biology of the Cell*. New York: Garland Science.

Alchourrón, C. (1993). Philosophical foundations of deontic logic and the logic of defeasible conditionals. In J.-J. Meyer & R. J. Wieringa (Eds.), *Deontic Logic in Computer Science: Normative System Specification* (pp. 43–84). Chichester: Wiley.

Alloway, T. P. (2009). Working memory, but not IQ, predicts subsequent learning in children with learning difficulties. *European Journal of Psychological Assessment*, 25(2), 92–8.

Allport, G. W. (2000). *Becoming: Basic Considerations for a Psychology of Personality*. New Haven, CT: Yale University Press. (Original work published 1955).

Andersson, G. (1994). *Criticism and the History of Science: Kuhn's, Lakatos's and Feyerabend's Criticism of Critical Rationalism*. Leiden: E. J. Brill.

Armstrong, D. M. (1983). *What Is a Law of Nature?* Cambridge: Cambridge University Press.

Armstrong, D. M. (1989). *Universals: An Opinionated Introduction*. Boulder, CO: Westview Press.

Baars, B. J. (1986). *The Cognitive Revolution in Psychology*. New York: Guilford Press.

Bachelard, G. (1974). *O Novo Espírito Científico*. São Paulo: Abril Cultural.

Baddeley, A. D., Gathercole, S. E., & Papagno, C. (1998). The phonological loop as a language learning device. *Psychological Review*, 105(1), 158–73.

Baddeley, A. D., & Hitch, G. J. (1974). Working memory. In G. H. Bower (Ed.), *The Psychology of Learning and Motivation*, Vol. 8 (pp. 47–89). New York: Academic Press. https://doi.org/10.1016/S0079-7421(08)60452-1

Bandura, A. (1977). *Social Learning Theory*. Englewood Cliffs, NJ: Prentice-Hall.

Baumeister, R. F. (2024). *The Science of Free Will: Bridging Theory and Positive Psychology*. New York: Oxford University Press.

Bechtel, W. (2008). *Mental Mechanisms: Philosophical Perspectives on Cognitive Neurosciences*. New York: Routledge.

Bechtel, W. (2009). Looking down, around, and up: Mechanistic explanations in psychology. *Philosophical Psychology*, 22(5), 543–64.

Bechtel, W., & Abrahamsen, A. (2005). Explanation: A mechanist alternative. *Studies in History and Philosophy of Biological and Biomedical Sciences*, 36(2), 421–41.

Bechtel, W., & Wright, C. (2009). What is psychological explanation? In P. Calvo & J. Symons (Eds.), *The Routledge Companion to the Philosophy of Psychology* (pp. 113–30). New York: Routledge.

Bird, A. (2005). The dispositionalist conception of laws. *Synthese*, 144(3), 325–55.

Bird, A. (2007). *Nature's Metaphysics: Laws and Properties*. Oxford: Oxford University Press.

Bliss, T. V. P., & Lømo, T. (1973). Long-lasting potentiation of synaptic transmission in the dentate area of the anaesthetized rabbit following stimulation of the perforant path. *Journal of Physiology*, 232(2), 331–56.

Bokulich, A. (2014). How the tiger bush got its stripes: "How possibly" vs. "how actually" model explanations. *The Monist*, 97(3), 321–38.
Brandon, R. N. (1990). *Adaptation and Environment*. Princeton, NJ: Princeton University Press.
Brentano, F. (1995). *Psychology from an Empirical Standpoint* (L. L. McAlister, Ed.; A. C. Rancurello, D. B. Terrell, & L. L. McAlister, Trans.). London: Routledge. (Original work published 1874).
Bronfenbrenner, U. (1979). *The Ecology of Human Development: Experiments by Nature and Design*. Cambridge, MA: Harvard University Press.
Bruner, J. S. (1990). *Acts of Meaning*. Cambridge, MA: Harvard University Press.
Carlsson, A., Lindqvist, M., Magnusson, T., & Waldeck, B. (1958). On the presence of 3-hydroxytyramine in brain. *Science (New York, N.Y.)*, 127(3296), 471. https://doi.org/10.1126/science.127.3296.471
Carrier, M. (1998). In defense of psychological laws. *International Studies in the Philosophy of Science*, 12(3), 217–32.
Cartwright, N. (1980). The truth doesn't explain much. *American Philosophical Quarterly*, 17(2), 159–63.
Cartwright, N. (1983). *How the Laws of Physics Lie*. Oxford: Oxford University Press.
Cartwright, N. (1989). *Nature's Capacities and Their Measurement*. Oxford: Clarendon Press.
Cartwright, N. (1999). *The Dappled World: A Study of the Boundaries of Science*. Cambridge: Cambridge University Press.
Cartwright, N. (2002). *From Causation to Explanation and Back. Technical Report*. London: Centre for Philosophy of Natural and Social Science, London School of Economics.
Cartwright, N. (2021). *Nature, the Artful Modeler: Lectures on Laws, Science, How Nature Arranges the World and How We Can Arrange It Better*. Chicago, IL: Open Court.
Case, R. (1998). The development of conceptual structures. In W. Damon (Series Ed.) & D. Kuhn & R. S. Siegler (Vol. Eds.), *Handbook of Child Psychology: Vol. 2. Cognition, Perception, and Language* (5th ed., pp. 745–800). New York: Wiley.
Castañon, G. (2007). *O que é Cognitivismo*. São Paulo: EPU.
Castañon, G. (2009). Psicologia como ciência moderna: vetos históricos e status atual. *Temas em Psicologia*, 17(1), 21–36.
Castañon, G. (2021a). Possibilist explanation: Explaining how-possibly through laws. *Erkenntnis*, 86, 835–52. https://doi.org/10.1007/s10670-019-00134-1.
Castañon, G. (2021b). *Conhecimento Possibilista* (Doctoral dissertation). Universidade Federal do Rio de Janeiro, Brazil.
Chomsky, N. (1980). *Rules and Representations*. New York: Columbia University Press.
Chomsky, N. (1995). *The Minimalist Program*. Cambridge, MA: MIT Press.
Churchland, P. M. (1981). Eliminative materialism and propositional attitudes. *The Journal of Philosophy*, 78(1), 67–90. https://doi.org/10.2307/2025900.
Cohen, S., & Greenberg, M. E. (2008). Communication between the synapse and the nucleus in neuronal development, plasticity, and disease. *Annual Review of Cell and Developmental Biology*, 24, 183–209. https://doi.org/10.1146/annurev.cellbio.24.110707.175235.
Collingwood, R. G. (1940). *An Essay on Metaphysics*. Oxford: Clarendon Press.
Comte, A. (1973). *Curso de Filosofia Positiva*. São Paulo: Abril Cultural.

Conway, A. R. A., Cowan, N., & Bunting, M. F. (2001). The cocktail party phenomenon revisited: The importance of working memory capacity. *Psychonomic Bulletin & Review*, 8(2), 331–5.

Craik, F. I. M., & Lockhart, R. S. (1972). Levels of processing: A framework for memory research. *Journal of Verbal Learning and Verbal Behavior*, 11(6), 671–84. https://doi.org/10.1016/S0022-5371(72)80001-X.

Craske, M. G., Treanor, M., Conway, C. C., Zbozinek, T., & Vervliet, B. (2014). Maximizing exposure therapy: An inhibitory learning approach. *Behaviour Research and Therapy*, 58, 10–23. https://doi.org/10.1016/j.brat.2014.04.006.

Craver, C. F. (2007). *Explaining the Brain: Mechanisms and the Mosaic Unity of Neuroscience*. Oxford: Oxford University Press.

Craver, C. F., & Tabery, J. (2023). Mechanisms in science. In E. N. Zalta (Ed.), *The Stanford Encyclopedia of Philosophy* (Fall 2023 Edition). https://plato.stanford.edu/archives/fall2023/entries/science-mechanisms/.

Csikszentmihalyi, M. (1990). *Flow: The Psychology of Optimal Experience*. New York: Harper & Row.

Cuffaro, M. E. (2015). How-possibly explanations in (quantum) computer science. *Philosophy of Science*, 82(5), 737–48. https://doi.org/10.1086/683144.

Cummins, R. (1983). *The Nature of Psychological Explanation*. Cambridge, MA: MIT Press.

Cummins, R. (2000). "How does it work?" versus "what are the laws?": Two conceptions of psychological explanation. In F. C. Keil & R. A. Wilson (Eds.), *Explanation and cognition* (pp. 117–45). Cambridge, MA: MIT Press.

Cummins, R. (2010). "How does it work?" vs. "What are the laws?" Two conceptions of psychological explanation. In *The World in the Head* (pp. 282–310). Oxford: Oxford University Press.

Dale, R., Chemero, A., & Dietrich, E. (2009). Explanatory pluralism in cognitive science. *Cognitive Science*, 33(5), 739–742. https://doi.org/10.1111/j.1551-6709.2009.01038.x.

Daly, H. E. (1996). *Beyond Growth: The Economics of Sustainable Development*. Boston, MA: Beacon Press.

Davidson, D. (1963). Actions, reasons, and causes. *The Journal of Philosophy*, 60(23), 685–700. https://doi.org/10.2307/2023177.

Davidson, D. (1980). Mental events. In N. Block (Ed.), *Readings in Philosophy of Psychology* (Vol. 1, pp. 107–19). Cambridge, MA: Harvard University Press.

De Carvalho, R. (1990). A history of the third force in psychology. *Journal of Humanistic Psychology*, 30(1), 22–44. https://doi.org/10.1177/0022167890301004.

Dennett, D. C. (1984). *Elbow Room: The Varieties of Free Will Worth Wanting*. Cambridge, MA: MIT Press.

Dennett, D. C. (1991). *Consciousness Explained*. Boston, MA: Little, Brown.

Descartes, R. (1952). *Meditations on First Philosophy* (1st ed. 1641). In R. M. Hutchins (Ed.), *Great Books of the Western World*, Vol. 31. Chicago: Encyclopaedia Britannica.

Descartes, R. (1984). Principles of Philosophy (1st ed. 1644). In J. Cottingham, R. Stoothoff & D. Murdoch (Trans.), *The Philosophical Works of Descartes* (Vol. 1, pp. 195–292). Cambridge: Cambridge University Press.

Dietz, S. (1970). Discussion: A remark on Hempel's replies to his critics. *Philosophy of Science*, 37(4), 614–17.

Dilthey, W. (1945). *Psicología y teoría del conocimiento* (E. Imaz, Trans.). Mexico City: Fondo de Cultura Económica.
Dowe, P. (1992). Wesley Salmon's process theory of causality and the conserved quantity theory. *Philosophy of Science*, 59(2), 195–216.
Dray, W. (1954). Explanatory narrative in history. *The Philosophical Quarterly*, 4(14), 15–27.
Dray, W. (1957). *Laws and Explanation in History*. Oxford: Clarendon Press.
Dray, W. (1968). On explaining how-possibly. *The Monist*, 52(3), 390–407.
Dreyfus, H. L. (1993). *What Computers Still Can't Do: A Critique of Artificial Reason*. Cambridge, MA: MIT Press.
Duhem, P. (1954). *The Aim and Structure of Physical Theory* (P. P. Wiener, Trans.). Princeton, NJ: Princeton University Press. (Original work published 1906).
Durkheim, É. (1951). *Suicide: A Study in Sociology* (J. A. Spaulding & G. Simpson, Trans.). Glencoe, IL: Free Press. (Original work published 1897).
Everett, D. L. (2005). Cultural constraints on grammar and cognition in Pirahã: Another look at the design features of human language. *Current Anthropology*, 46(4), 621–46. https://doi.org/10.1086/431525.
Everett, D. L. (2012). *Language: The Cultural Tool*. New York: Pantheon Books.
Febvre, L. (1932). *A Geographical Introduction to History*. London: Kegan Paul, Trench, Trubner.
Festinger, L. (1957). *A Theory of Cognitive Dissonance*. Stanford, CA: Stanford University Press.
Fischer, J. M. (1999). Recent work on moral responsibility. *Ethics*, 110(1), 91–139.
Fischer, J. M., & Tognazzini, N. A. (2013). The logic of theological incompatibilism: A reply to Westphal. *Analysis*, 73(1), 46–48.
Fodor, J. A. (1974). Special sciences (or: The disunity of science as a working hypothesis). *Synthese*, 28(2), 97–115.
Fodor, J. A. (1975). *The Language of Thought*. Cambridge, MA: Harvard University Press.
Fodor, J. A. (1991a). You can fool some of the people all of the time, everything else being equal; hedged laws and psychological explanation. *Mind*, 100(1), 19–34.
Fodor, J. A. (1991b). Methodological solipsism considered as a research strategy in cognitive psychology. In R. Boyd & P. Gasper (Eds.), *The Philosophy of Science* (pp. 651–69). Cambridge, MA: MIT Press.
Fodor, J. A. (1968). *Psychological Explanation: An Introduction to the Philosophy of Psychology*. New York: Random House.
Forber, P. (2010). Confirmation and explaining how possible. *Studies in History and Philosophy of Biological and Biomedical Sciences*, 41(1), 32–40. https://doi.org/10.1016/j.shpsc.2009.12.002.
Frankfurt, H. G. (1971). Freedom of the will and the concept of a person. *The Journal of Philosophy*, 68(1), 5–20.
Frankl, V. E. (2006). *Man's Search for Meaning*. Boston, MA: Beacon Press. (Original work published 1946).
Freud, S. (1949). *An Outline of Psychoanalysis* (J. Strachey, Trans.). New York: W. W. Norton. (Original work published 1940).
Gadenne, V. (2006). *Filosofía de la Psicología*. Barcelona: Herder Editorial.
Gaj, N. (2016). *Unity and Fragmentation in Psychology: The Philosophical and Methodological Roots of the Discipline*. London: Routledge.

Galilei, G. (1957). *The Assayer*. In S. Drake (Ed. & Trans.), *Discoveries and Opinions of Galileo* (pp. 231–280). New York: Doubleday. (Original work published 1623).

Gardner, H., & Laskin, E. (1995). *Leading Minds: An Anatomy of Leadership*. New York: Basic Books.

Gardner, H. (1992). Scientific psychology: Should we bury it or praise it? *New Ideas in Psychology*, 10(2), 179–90.

Gathercole, S. E., Pickering, S. J., Ambridge, B., & Wearing, H. (2004). The structure of working memory from 4 to 15 years of age. *Developmental Psychology*, 40(2), 177–90.

Georgescu-Roegen, N. (1971). *The Entropy Law and the Economic Process*. Cambridge, MA: Harvard University Press.

Gergen, K. J. (1973). Social psychology as history. *Journal of Personality and Social Psychology*, 26(2), 309–20.

Gergen, K. J. (1985). The social constructionist movement in modern psychology. *American Psychologist*, 40(3), 266–75.

Giere, R. N. (2006). *Scientific Perspectivism*. Chicago, IL: University of Chicago Press.

Gifford, R. (2014). *Environmental Psychology: Principles and Practice* (5th ed.). Colville, WA: Optimal Books.

Glennan, S. (2017). *The New Mechanical Philosophy*. Oxford: Oxford University Press.

Glymour, C. (2001). *The Mind's Arrows: Bayes Nets and Graphical Causal Models in Psychology*. Cambridge, MA: MIT Press.

Goertzen, J. R. (2008). On the possibility of unification: The reality and nature of the crisis in psychology. *Theory and Psychology*, 18(6), 829–52. https://doi.org/10.1177/0959354308097260.

Goertz, G. (2003). The substantive importance of necessary conditions hypothesis. In G. Goertz & H. Starr (Eds.), *Necessary Conditions: Theory, Methodology and Applications* (pp. 65–94). Lanham, MD: Rowman & Littlefield.

Goertz, G. (2012). Descriptive causal generalizations: Empirical laws in the social sciences? In H. Kincaid (Ed.), *The Oxford Handbook of Philosophy of Social Science* (pp. 85–107). New York: Oxford University Press.

Goertz, G., & Starr, H. (2003). *Necessary Conditions: Theory, Methodology and Applications*. Lanham, MD: Rowman & Littlefield.

Goodman, N. (1955). *Fact, Fiction, and Forecast*. Cambridge, MA: Harvard University Press.

Greenspan, P. S. (1978). Oughts and determinism: A response to Goldman. *The Philosophical Review*, 87(1), 77–83.

Grüne-Yanoff, T. (2009). Learning from minimal economic models. *Erkenntnis*, 70(1), 81–99.

Hale, B. (2013). *Necessary Beings: An Essay on Ontology, Modality, and the Relations Between Them*. Oxford: Oxford University Press.

Hart, H. L. A., & Honoré, A. M. (1959). *Causation in the Law*. Oxford: Oxford University Press.

Hart, H. L. A., & Honoré, A. M. (1985). *Causation in the Law* (2nd ed.). Oxford: Oxford University Press.

Hartline, D. K., & Colman, D. R. (2007). Rapid conduction and the evolution of giant axons and myelinated fibers. *Current Biology*, 17(1), R29–R35. https://doi.org/10.1016/j.cub.2006.11.042.

Hempel, C. G., & Oppenheim, P. (1948). Studies in the logic of explanation. *Philosophy of Science*, 15(2), 135–75.

Hempel, C. G. (1965). *Aspects of Scientific Explanation and Other Essays in the Philosophy of Science*. New York: Free Press.
Hempel, C. G. (1966). *Philosophy of Natural Science*. Englewood Cliffs, NJ: Prentice-Hall.
Henriques, G. R. (2011). *A New Unified Theory of Psychology*. New York: Springer.
Hitchcock, C. (1995). Discussion: Salmon on explanatory relevance. *Philosophy of Science*, 62(2), 304–20.
Hoff, E. (2006). How social contexts support and shape language development. *Developmental Review*, 26(1), 55–88.
Hubel, D. H., & Wiesel, T. N. (1962). Receptive fields, binocular interaction and functional architecture in the cat's visual cortex. *Journal of Physiology*, 160(1), 106–54.
Hume, D. (1952). *An Enquiry Concerning Human Understanding*. Oxford: Oxford University Press. (Original work published 1748).
Hurley, P. J. (2018). *A Concise Introduction to Logic* (13th ed.). Boston, MA: Cengage Learning.
Husserl, E. (1952). *A Filosofia como Ciência de Rigor*. Coimbra: Coimbra Editora. (Original work published 1911).
Johnston, R. J., Gregory, D., Pratt, G., & Watts, M. (2009). *The Dictionary of Human Geography* (5th ed.). Chichester: Wiley-Blackwell.
Jordan, Z. (1963). Logical determinism. *Notre Dame Journal of Formal Logic*, 4(1), 1–16.
Kaiser, M. I. (2012). *An Ontic Account of Explanatory Reduction in Biology* (Doctoral dissertation). University of Cologne.
Kandel, E. R. (1998). A new intellectual framework for psychiatry. *American Journal of Psychiatry*, 155(4), 457–69.
Kant, I. (2001). *Critique of Pure Reason* (P. Guyer & A. W. Wood, Trans.). Cambridge: Cambridge University Press. (Original work published 1781).
Kant, I. (2004). *Metaphysical Foundations of Natural Science* (M. Friedman, Trans. & Ed.). Stanford, CA: Stanford University Press. (Original work published 1786).
Kenrick, D. T., Griskevicius, V., Neuberg, S. L., & Schaller, M. (2010). Renovating the pyramid of needs: Contemporary extensions built upon ancient foundations. *Perspectives on Psychological Science*, 5(3), 292–314.
Kimble, G. A. (1984). Psychology's two cultures. *American Psychologist*, 39(8), 833–9.
Kitcher, P. (1989). Explanatory unification and the causal structure of the world. In P. Kitcher & W. C. Salmon (Eds.), *Scientific Explanation* (*Minnesota Studies in the Philosophy of Science*, Vol. 13, pp. 410–505). Minneapolis: University of Minnesota Press.
Kitcher, P. (1993). *The Advancement of Science: Science without Legend, Objectivity Without Illusions*. Oxford: Oxford University Press.
Kivipelto, M., Solomon, A., & Mangialasche, F. (2021). High-resolution structural MRI and its application in the assessment of hippocampal integrity and connectivity in cognitive impairment. *Brain Communications*, 3(2), fcab088. https://doi.org/10.1093/braincomms/fcab088.
Koch, S. (Ed.). (1959). *Psychology: A Study of a Science. Study I. Conceptual and Systematic: History and Systematics*. New York: McGraw-Hill.
Koch, S. (1993). Psychology or the psychological studies? *American Psychologist*, 48(8), 902–4.
Kornack, D. R., & Rakic, P. (2001). Cell proliferation without neurogenesis in adult primate neocortex. *Science*, 294(5549), 2127–30. https://doi.org/10.1126/science.1065467.

Kripke, S. A. (1980). *Naming and Necessity*. Cambridge, MA: Harvard University Press.

Kruger, J., & Dunning, D. (1999). Unskilled and unaware of it: How difficulties in recognizing one's own incompetence lead to inflated self-assessments. *Journal of Personality and Social Psychology*, 77(6), 1121–34.

Kuhl, P. K. (2010). Brain mechanisms in early language acquisition. *Neuron*, 67(5), 713–27.

Kuhn, T. S. (1962). *The Structure of Scientific Revolutions*. Chicago, IL: University of Chicago Press.

La Blache, P. V. de. (1908). *La France: Tableau géographique*. Paris: Hachette.

La Blache, P. V. (1967). *Principles of Human Geography* (E. de Martonne, Ed.; M. T. Bingham, Trans.). London: Constable.

Leibniz, G. W. (1974). *Novos Ensaios sobre o Entendimento Humano*. São Paulo: Abril Cultural. (Original work published 1704).

Leibniz, G. W. (1989). Principles of nature and grace, based on reason. In R. Ariew & D. Garber (Eds. & Trans.), *Philosophical Essays* (p. 210). Indianapolis, IN: Hackett. (Original work published 1714).

Leibniz, G. W. (1998). Specimen dynamicum: An essay in dynamics. In R. Francks & R. S. Woolhouse (Trans. & Eds.), *Philosophical Texts* (pp. 153–79). Oxford: Oxford University Press. (Original work published 1695).

Leite, D. A. (2021). *The Twenty-First Century Mechanistic Theory of Human Cognition: A Critical Analysis*. Cham, Switzerland: Springer Nature. https://doi.org/10.1007/978-3-030-63680-7.

Levy, J., & Goertz, G. (2007). *Explaining War and Peace: Case Studies and Necessary Condition Counterfactuals*. London: Routledge.

Lewis, D. (1973a). Causation. *The Journal of Philosophy*, 70(17), 556–67.

Lewis, D. (1973b). *Counterfactuals*. Cambridge, MA: Harvard University Press.

Lewis, D. (1983). New work for a theory of universals. *Australasian Journal of Philosophy*, 61(4), 343–77.

Lewis, D. (1986). *On the Plurality of Worlds*. Oxford: Blackwell.

Libet, B., Gleason, C. A., Wright, E. W., & Pearl, D. K. (1983). Time of conscious intention to act in relation to onset of cerebral activity (readiness-potential): The unconscious initiation of a freely voluntary act. *Brain*, 106(3), 623–42.

Locke, J. (1978). *Concerning Human Understanding*. In *Britannica Great Books*, Vol. 35. Chicago, IL: Encyclopaedia Britannica. (Original work published 1689).

Longino, H. E. (1990). *Science as Social Knowledge: Values and Objectivity in Scientific Inquiry*. Princeton, NJ: Princeton University Press.

Lu, B. (2003). BDNF and its role in synaptic plasticity and neurogenesis. *Nature Reviews Neuroscience*, 4(11), 839–49. https://doi.org/10.1038/nrn1204.

Luce, R. D. (1986). *Response Times: Their Role in Inferring Elementary Mental Organization*. New York: Oxford University Press.

Machamer, P., Darden, L., & Craver, C. F. (2000). Thinking about mechanisms. *Philosophy of Science*, 67(1), 1–25.

Mackie, J. L. (1965). Causes and conditions. *American Philosophical Quarterly*, 2(4), 245–64.

Mackie, J. L. (1980). *The Cement of the Universe: A Study of Causation*. Oxford: Oxford University Press.

Malinowski, B. (1936). Anthropology. In *Encyclopaedia Britannica*, Supplement Vol. I. London: Encyclopaedia Britannica.
Malthus, T. R. (1998). *An Essay on the Principle of Population* (G. Gilbert, Ed.). Oxford: Oxford University Press. (Original work published 1798).
Manero, J. J. (2023). Rights as formal combinations of normative variables. *Revus – Journal for Constitutional Theory and Philosophy of Law*, 59, 65–98. https://doi.org/10.3986/Revus.2023.59.4.
Marr, D. (1982). *Vision: A Computational Investigation into the Human Representation and Processing of Visual Information*. San Francisco, CA: W. H. Freeman.
Maslow, A. H. (1954). *Motivation and Personality*. New York: Harper & Row.
Mazzoni, J. (2018). *Fisicalismo: O materialismo contemporâneo sem matéria?* [Doctoral thesis, Universidade Federal de Juiz de Fora].
Meehl, P. E. (1978). Theoretical risks and tabular asterisks: Sir Karl, Sir Ronald, and the slow progress of soft psychology. *Journal of Consulting and Clinical Psychology*, 46(4), 806–34.
Meinong, A. (1904). Über Gegenstandstheorie. In A. Meinong (Ed.), *Untersuchungen zur Gegenstandstheorie und Psychologie* (pp. 1–50). Leipzig: Barth.
Merlussi, P. (2017). *Laws of Nature and Free Will* (Doctoral dissertation). Durham University. http://etheses.dur.ac.uk/12469/.
Merlussi, P., & Cartwright, N. (2018). Are laws of nature consistent with contingency? In W. Ott & L. Patton (Eds.), *Laws of Nature*. Oxford: Oxford University Press.
Mumford, S. (1998). *Dispositions*. Oxford: Oxford University Press.
Mumford, S., & Anjum, R. L. (2011). *Getting Causes from Powers*. Oxford: Oxford University Press.
Mumford, S., & Anjum, R. L. (2014). A new argument against compatibilism. *Analysis*, 74(1), 20–25.
Mumford, S., & Anjum, R. L. (2015). Freedom and control: On the modality of free will. *American Philosophical Quarterly*, 52(1), 1–12.
Murdock, B. B., Jr. (1962). The serial position effect of free recall. *Journal of Experimental Psychology*, 64(5), 482–8.
Nagel, E. (1961). *The Structure of Science: Problems in the Logic of Scientific Explanation*. New York: Harcourt, Brace & World.
Neisser, U. (1967). *Cognitive Psychology*. New York: Appleton-Century-Crofts.
Neisser, U. (1975). *Cognition and Reality: Principles and Implications of Cognitive Psychology*. San Francisco, CA: W. H. Freeman.
Nemeroff, C. B. (2003). The role of GABA in the pathophysiology and treatment of anxiety disorders. In D. J. Stein & E. Hollander (Eds.), Textbook of anxiety disorders (pp. 131–142). Washington, DC: American Psychiatric Publishing.
Nielsen, M. A. & Chuang, I. L. (2011). *Quantum computation and quantum information* (10th anniversary ed.). Cambridge University Press.
Niiniluoto, I. (1999). *Critical Scientific Realism*. Oxford: Oxford University Press.
O"Hara, R. J. (1988). Homage to Clio, or, toward a historical philosophy for evolutionary biology. *Systematic Zoology*, 37(2), 142–55.
Passmore, J. (1958). Review of *Law and Explanation in History*, by William Dray. *Australian Journal of Politics and History*, 4(2), 269–75.
Pavlov, I. P. (2010). *Conditioned Reflexes: An Investigation of the Physiological Activity of the Cerebral Cortex* (G. V. Anrep, Trans.). Mineola, NY: Dover Publications. (Original work published 1927).

Paller, K. A., & Wagner, A. D. (2002). Observing the transformation of experience into memory. *Trends in Cognitive Sciences*, 6(2), 93–102. https://doi.org/10.1016/S1364-6613(00)01845-3.
Penna, A. G. (1991). *História das Idéias Psicológicas*. Rio de Janeiro: Imago.
Persson, J. (2009). Mechanistic explanation in social contexts: Elster and the problem of local scientific growth. *Philosophical Explorations*, 12(1), 81–96. https://doi.org/10.1080/13869790802635867.
Persson, J. (2012). Three conceptions of explaining how possibly—and one reductive account. In H. de Regt, S. Hartmann, & S. Okasha (Eds.), *EPSA Philosophy of Science: Amsterdam 2009* (The European Philosophy of Science Association Proceedings, Vol. 1, pp. 227–37). Dordrecht: Springer.
Piaget, J., & Inhelder, B. (1941). *Le Développement des Quantités chez l'Enfant*. Neuchâtel: Delachaux et Niestlé.
Piaget, J. (1965). *The Child's Conception of Number*. New York: W. W. Norton.
Pnueli, A. (1977). The temporal logic of programs. *Proceedings of the 18th Annual Symposium on Foundations of Computer Science*, 46–57.
Popper, K. R. (1979). *Objective Knowledge* (Rev. ed.). Oxford: Clarendon Press.
Popper, K. R. (2002). *The Logic of Scientific Discovery*. London: Routledge. (Original work published 1959).
Popper, K. R., & Eccles, J. C. (1977). *The Self and Its Brain: An Argument for Interactionism*. Berlin: Springer.
Putnam, H. (1961). Minds and machines. In S. Hook (Ed.), *Dimensions of Mind* (pp. 221–31). New York: Collier.
Pylyshyn, Z. W. (1984). *Computation and Cognition: Toward a Foundation for Cognitive Science*. Cambridge, MA: MIT Press.
Quine, W. V. O. (1948). On what there is. *The Review of Metaphysics*, 2(5), 21–38.
Quine, W. V. O. (1951). Two dogmas of empiricism. *The Philosophical Review*, 60(1), 20–43.
Railton, P. (1978). A deductive-nomological model of probabilistic explanation. *Philosophy of Science*, 45(2), 206–26.
Reichenbach, H. (1947). *Elements of Symbolic Logic*. New York: Macmillan.
Reiner, R. (1993). Necessary conditions and explaining how-possibly. *Philosophical Quarterly*, 43(170), 58–69.
Resnik, D. B. (1991). How-possibly explanations in biology. *Acta Biotheoretica*, 39(2), 141–9.
Reydon, T. A. C. (2012). How-possibly explanations as genuine explanations and helpful heuristics: A comment on Forber. *Studies in History and Philosophy of Biological and Biomedical Sciences*, 43(1), 302–10.
Robinson, D. N. (1985). *Philosophy of Psychology*. New York: Columbia University Press.
Rock, I. (1984). *Perception*. New York: Scientific American Library.
Roese, N. J., & Olson, J. M. (1995). *What Might Have Been: The Social Psychology of Counterfactual Thinking*. Hillsdale, NJ: Lawrence Erlbaum Associates.
Rosbash, M., & Hall, J. C. (1989). The molecular biology of circadian rhythms. *Neuron*, 3(5), 387–98. https://doi.org/10.1016/0896-6273(89)90014-4.
Rosenberg, A. (1996). Laws, damn laws, and ceteris paribus clauses. *The Southern Journal of Philosophy*, 34(S1), 183–204.

Rosenberg, A. (2011). *Philosophy of Science: A Contemporary Introduction*. New York: Routledge.

Rosenberg, A. (2012). *Philosophy of Social Science* (4th ed.). Boulder, CO: Westview Press.

Royce, J. R. (1985). The problem of theoretical pluralism in psychology. In K. B. Madsen & L. P. Mos (Eds.), *Annals of Theoretical Psychology* (Vol. 3, pp. 297–311). New York: Plenum Press.

Rupert, R. D. (2008). Ceteris paribus laws, component forces, and the nature of special-science properties. *Noûs*, 42(3), 349–80.

Ryan, R. M., & Deci, E. L. (2000). Self-determination theory and the facilitation of intrinsic motivation, social development, and well-being. *American Psychologist*, 55(1), 68–78. https://doi.org/10.1037/0003-066X.55.1.68.

Rychlak, J. F. (1975). Psychological science as a humanist views it. *Nebraska Symposium on Motivation*, 23, 205–79.

Rychlak, J. F. (1993). A suggested principle of complementarity for psychology: In theory, not method. *American Psychologist*, 48(9), 933–42.

Rychlak, J. F. (1994). *Logical Learning Theory: A Human Teleology and Its Empirical Support*. Lincoln: University of Nebraska Press.

Salmon, W. C. (1971). Statistical explanation. In W. C. Salmon (Ed.), *Statistical Explanation and Statistical Relevance* (pp. 29–87). Pittsburgh, PA: University of Pittsburgh Press.

Salmon, W. C. (1984). *Scientific Explanation and the Causal Structure of the World*. Princeton, NJ: Princeton University Press.

Salmon, W. C. (1989). *Four Decades of Scientific Explanation*. Pittsburgh, PA: University of Pittsburgh Press.

Schiffer, S. R. (1991). Ceteris paribus laws. *Mind*, 100(1), 1–17.

Scoville, W. B., & Milner, B. (1957). Loss of recent memory after bilateral hippocampal lesions. *Journal of Neurology, Neurosurgery and Psychiatry*, 20(1), 11–20.

Searle, J. R. (1958). Proper names. *Mind*, 67(266), 166–73. Reprinted in P. F. Strawson (Ed.), *Philosophical Logic* (1967, pp. 89–96). Oxford: Oxford University Press.

Searle, J. R. (1992). *The Rediscovery of the Mind*. Cambridge, MA: MIT Press.

Searle, J. R. (2001). *Rationality in Action*. Cambridge, MA: MIT Press.

Silva, L. (2024). *O problema da unidade da psicologia e a possibilidade de sua unificação teórica* (Doctoral dissertation). Universidade Federal de Juiz de Fora, Brazil.

Skinner, B. F. (1938). *The Behavior of Organisms: An Experimental Analysis*. New York: Appleton-Century.

Skinner, B. F. (1953). *Science and Human Behavior*. New York: Macmillan.

Skocpol, T. (1979). *States and Social Revolutions: A Comparative Analysis of France, Russia and China*. Cambridge: Cambridge University Press.

Spelke, E. S. (2000). Core knowledge. *American Psychologist*, 55(11), 1233–43.

Staats, A. W. (1983). *Psychology's Crisis of Disunity: Philosophy and Method for a Unified Science*. New York: Praeger.

Staats, A. W. (1991). Unified positivism and unification psychology: Fad or new field? *American Psychologist*, 46(9), 899–912.

Sternberg, R. J. (Ed.). (2004). *Unity in Psychology: Possibility or Pipedream?* Washington, DC: American Psychological Association.

Sternberg, R. J. (2004). Unifying the field of psychology. In R. J. Sternberg (Ed.), *Unity in Psychology: Possibility or Pipedream?* (pp. 3–14). Washington, DC: American Psychological Association.

Strawson, P. F. (1959). Review of *Law and Explanation in History*, by William Dray. *Mind*, 68(270), 265–8.
Swanson, H. L., Zheng, X., & Jerman, O. (2009). Working memory, short-term memory, and reading disabilities: A selective meta-analysis of the literature. *Journal of Learning Disabilities*, 42(3), 260–87.
Terrace, H. S., Petitto, L. A., Sanders, R. J., & Bever, T. G. (1979). Can an ape create a sentence? *Science*, 206(4421), 891–902. https://doi.org/10.1126/science.504995.
Thagard, P. (2003). Pathways to biomedical discovery. *Philosophy of Science*, 70(2), 235–54.
The World. (2016, July 30). Robert Bales pleads guilty to massacre of Afghan civilians, avoiding the death penalty. *The World*. https://theworld.org/stories/2016/07/30/robert-bales-pleads-guilty-massacre-afghan-civilians-avoiding-death-penalty.
Timmerman, T., & Cohen, Y. (2020). Actualism and possibilism in ethics. In E. N. Zalta (Ed.), *The Stanford Encyclopedia of Philosophy* (Fall 2020 Edition). Metaphysics Research Lab, Stanford University. https://plato.stanford.edu/archives/fall2020/entries/actualism-possibilism-ethics/.
Van Fraassen, B. C. (1980). *The Scientific Image*. Oxford: Clarendon Press.
Van Inwagen, P. (1983). *An Essay on Free Will*. Oxford: Oxford University Press.
Van Inwagen, P. (2008). The consequence argument. In P. Van Inwagen & D. W. Zimmerman (Eds.), *Metaphysics: The Big Questions* (pp. 309–17). Oxford: Blackwell.
Varma, S. (2025). *The Science of Free Will: How Determinism Affects Everything from the Future of AI to Traffic to God to Bees*. London: Iff Books.
Vihvelin, K. (2013). *Causes, Laws, and Free Will: Why Determinism Doesn't Matter*. Oxford: Oxford University Press.
Vihvelin, K. (2015). How not to think about free will. *Journal of Cognition and Neuroethics*, 3(1), 393–403.
Wahba, M. A., & Bridwell, L. G. (1976). Maslow reconsidered: A review of research on the need hierarchy theory. *Organizational Behavior and Human Performance*, 15(2), 212–40.
Waxman, S. G. (1980). Determinants of conduction velocity in myelinated nerve fibers. *Muscle & Nerve*, 3(2), 141–50. https://doi.org/10.1002/mus.880030207.
Weiskopf, D. A. (2011). Models and mechanisms in psychological explanation. *Synthese*, 183(3), 313–38.
Woodward, J. (1989). The causal mechanical model of explanation. *Minnesota Studies in the Philosophy of Science*, 13, 357–83.
Woodward, J. (2000). Explanation and invariance in the special sciences. *British Journal for the Philosophy of Science*, 51(2), 197–254.
Woodward, J. (2002). There is no such thing as a ceteris paribus law. In J. Earman, C. Glymour, & S. Mitchell (Eds.), *Erkenntnis*, 52(Special Issue), 303–28.
Woodward, J. (2003). *Making Things Happen: A Theory of Causal Explanation*. Oxford: Oxford University Press.
Woodward, J. (2019). Scientific explanation. In E. N. Zalta (Ed.), *The Stanford Encyclopedia of Philosophy* (Winter 2019 Edition). https://plato.stanford.edu/archives/win2019/entries/scientific-explanation/
Woodward, J., & Hitchcock, C. (2003). Explanatory generalizations, Part I: A counterfactual account. *Nous*, 37(1), 1–24.
Wootters, W. K., & Zurek, W. H. (1982). A single quantum cannot be cloned. *Nature*, 299(5886), 802–3. https://doi.org/10.1038/299802a0.

Wright, C. (2007). Reductionism in the mind sciences. In M. Matthen & C. Stephens (Eds.), *Philosophy of Biology* (pp. 387–409). Amsterdam: Elsevier.

Wright, C., & Bechtel, W. (2007). Mechanisms and psychological explanation. In P. Thagard (Ed.), *Philosophy of Psychology and Cognitive Science* (pp. 31–77). Amsterdam: Elsevier.

Yanchar, S. C., & Slife, B. D. (1997). Pursuing unity in a fragmented psychology: Problems and prospects. *Review of General Psychology*, 1(3), 235–55.

Zagare, F. (1996). Classical deterrence theory: A critical assessment. *International Interactions*, 21: 365–88.

Zeng, H., Huang, J., Jin, Z., & Ma, H. (2021). Hippocampal subfields volume changes and its impact on memory function in Alzheimer's disease. *Neurobiology of Aging*, 56, 123–30.

Zotey, V., Andhale, A., Shegekar, T., et al. (2023). Adaptive neuroplasticity in brain injury recovery: Strategies and insights. *Cureus*, 15(9), e45873.

Index

www.ingramcontent.com/pod-product-compliance
Ingram Content Group UK Ltd.
Pitfield, Milton Keynes, MK11 3LW, UK
UKHW020216290926
15470UKWH00001B/229

9 781666 95396